Earthquake Design of Concrete Masonry Buildings:

Strength Design of One- to Four-Story Buildings

Earthquake Design of Concrete Masonry Buildings:

Volume 2

Strength Design of One- to Four-Story Buildings

Robert E. Englekirk

Gary C. Hart

*Englekirk and Hart
Consulting Engineers, Inc.*

and

The Concrete Masonry Association
of California and Nevada

PRENTICE-HALL, INC., Englewood Cliffs, New Jersey 07632

Library of Congress Cataloging in Publication Data
(Revised for vol. 2)

Englekirk, Robert E.,
    Earthquake design of concrete masonry buildings.

    Includes bibliographical references and index.
    Contents: v. 1. Response spectra analysis and general earthquake modeling—v. 2. Strength
design of one- to four-story buildings.
    1. Earthquakes and building. 2. Concrete construction.
3. Masonry. I. Hart, Gary C. II. Concrete Masonry
Association of California and Nevada. III. Title.
TH1095.E53   693.8′52   81-21141
ISBN 0-13-223065-8 (v. 1)
ISBN 0-13-223156-5 (v. 2)

Editorial/production supervision: Mary Carnis
Manufacturing buyer: Anthony Caruso

© 1984 by Prentice-Hall, Inc., Englewood Cliffs, New Jersey 07632

Printed in the United States of America

10   9   8   7   6   5   4   3   2   1

ISBN   0-13-223156-5

Prentice-Hall International, Inc., *London*
Prentice-Hall of Australia Pty. Limited, *Sydney*
Editora Prentice-Hall do Brasil, Ltda., *Rio de Janeiro*
Prentice-Hall Canada Inc., *Toronto*
Prentice-Hall of India Private Limited, *New Delhi*
Prentice-Hall of Japan, Inc., *Tokyo*
Prentice-Hall of Southeast Asia Pte. Ltd., *Singapore*
Whitehall Books Limited, *Wellington, New Zealand*

with a special thank you to

STUART BEAVERS

CONTENTS

PREFACE

The Concrete Masonry Association of California and Nevada recognizes that the safe and economical design of buildings often requires the structural engineer to perform an earthquake design that is state-of-the-art. Therefore, the Association has committed funding for the development of a three-book series on the earthquake design of concrete masonry buildings. Robert E. Englekirk and Gary C. Hart are writing the three books. Stuart Beavers is coordinating the efforts of the Concrete Masonry Association of California and Nevada.

This is the second book in the earthquake design series. The first book provided a foundation by discussing such topics as earthquake response spectra, static earthquake design forces, modeling of shear wall buildings, ductility, modes of vibration, and the geotechnical consultant's role in earthquake design. This book will go into depth in the design process for one- to four-story concrete masonry buildings. The third book will discuss earthquake design procedures for concrete masonry buildings of irregular shape or stiffness and buildings taller than four stories.

Only recently have U.S. colleges and universities begun to include structural dynamics and masonry building design in their civil engineering curricula. Thus, two groups of structural engineers will be assisted by this text. The first group is composed of professional engineers currently involved in structural design who wish to educate themselves in earthquake design; the second group is comprised of college students who `require such material on earthquake design procedures.

The authors wish to express their gratitude to Thomas A. Sabol and Sampson C. Huang for their efforts in helping us to clarify our ideas and improve the presentation of the material in this book. We also wish to acknowledge Dave Nickerson of KORFIL and Bruce Wilcox of the Berkeley Solar Group for their assistance in writing Section 2.7 on energy transmission.

Robert E. Englekirk
Gary C. Hart

Earthquake Design of Concrete Masonry Buildings:

Strength Design of One- to Four-Story Buildings

1

INTRODUCTION

1.1 GENERAL

Today's consumers require that products developed for their use not only be safe but economical as well. Concrete masonry buildings must then attempt to attain a proper balance between safety and economy. The professional engineer charged with the design of a building must be equipped to successfully execute such a design, and it is the goal of this book to provide the reader with the tools and information required to design concrete masonry buildings.

Historically, concrete masonry buildings have been designed using the working stress approach. Both the current *Uniform Building Code* (UBC) [1.1] and American Concrete Institute (ACI) [1.2] concrete masonry design standards are based on professional engineering experience and the working stress theories of structural mechanics. This book presents an alternative to such a design approach. The alternative is a two-level design approach in which one level addresses serviceability and the second level addresses strength. Such a design approach is logical and can provide significant savings in construction costs because the design process is more focused. The two-level design process provides the engineer with a more consistent relationship between loads and material capacities, and clearly separates serviceability considerations from strength considerations so that the professional can determine the impact of each on a specific building.

1.2 CONCEPTUAL DESIGN

The true design process is one which addresses the function of the building and selects the most economical means of satisfying these functional needs. Structural economy is usually viewed in terms of construction or development cost only. The designer must also consider the effect of serviceability of the structure on the long-term cost. To this end, a two-level design procedure which considers not only strength but serviceability is essential.

Imagine that an engineer is retained to design an industrial building. The first step in the design process is the *conceptual design* step. Here the engineer must: (1) select the structural system, (2) formulate a loading criterion, and (3) establish the functional constraints. Let us take each of these separately.

The *structural system* must be selected to satisfy the functional needs of the user as well as the loads imposed by environmental conditions. Consequently, the system must provide not only a continuous load path from the point of application of loads to the foundation but also consider the space planning needs and economic constraints of the user. Building systems and most building codes are developed to produce typical low-rise buildings which are safe, serviceable, and economical. Buildings which are of irregular plan geometry may, however, require specialized design considerations.

The way in which the designer considers the *loading conditions* acting on the structural system presents the first major departure of the two-level design approach from the existing working stress standards. Consider, for example, the wind loads. American National Standards Institute (ANSI) A58.1 (1982) standard develops a wind load that is a function of surface exposure, shape of structure, geographic location, and other parameters [1.3]. The design wind load is obtained by multiplying a load factor times a wind load, W, that is expected to occur on the average of once every 148 years. A reasonable serviceability design load might be $1.0W$ (i.e., a unit value for the load factor), while a load of $1.3W$ might be chosen to represent the strength design load.

Major developments have been made in the last two decades in understanding the magnitude/frequency relationship of all loads, and it is possible for the ANSI A58.1 (1982) loads to be interpreted from a magnitude/frequency-of-occurrence perspective. For example, it can be shown that if one uses a serviceability wind design load of $1.0W$, it will occur on the average of once every 148 years. It will also have approximately a 19 percent chance of being exceeded in 50 years. Similarly, a strength design load of $1.3W$ will occur *on the average* approximately once every 1061 years! It still has a small, approximately 5 percent, chance of occurring in the next 50 years. When loading is viewed from a service load/strength load (i.e., two-level) perspective, it helps us to understand better the loadings themselves and to communicate more clearly this area of our training and expertise to our clients. The latter benefit can be critically important because such input often aids clients in making decisions that maximize the long-term economy of design instead of just minimizing the immediate short-term costs. For example, discussions with a client regarding the economic impact of serviceability load selection on the short-term and long-term costs of a project may justify using serviceability loads that are either higher or lower than those used in traditional design projects.

The *functional constraints* are usually developed from the serviceability design load criterion. A functional constraint might be the maintenance of a watertight wall. The load-induced moment to satisfy this functional requirement might be one which is less than the cracking moment for the materials. Building codes should provide guidelines to assist the engineer in developing acceptable functional constraints for serviceability and strength. For complex projects, this may involve a component-by-component prescription.

Structural system, loading conditions, and functional constraints are the three integrated parts of a good conceptual design. Unfortunately, the engineer usually is not retained to explore thoroughly the possible permutations of the factors affecting these three parts. The construction cost and design fee in most building projects is such that the building is designed based on solutions shown to be acceptable for typical buildings. It is in this context that a code or standard provides a valuable aid to the engineer.

1.3 DESIGN OF ONE- TO FOUR-STORY BUILDINGS

In the development of this three-volume design series, the division of scope seemed to divide itself naturally into general earthquake engineering (Volume 1), low-rise or more typical buildings where code design is well within the state-of-the-practice (Volume 2), and high-rise or major-design-fee buildings where state-of-the-practice design requires dynamic analyses involving computer models (Volume 3).

The material contained in Volume 1 should satisfy the reader's desires relative to earthquake engineering [1.4]. In the development of this book, we discovered that to present a solid treatment of low-rise building design, it was necessary to develop a textbook in concrete masonry design because no books adequately address the behavior of concrete masonry in a manner consistent with structural mechanics. Therefore, this volume is a basic textbook in concrete masonry design *and* a design book for low-rise buildings. The term "low-rise" is not well defined and we chose to address one- to four-story buildings because the UBC makes a clear break in many areas at the four-story level.

Figure 1.1 shows the earthquake design process as we visualize it. Because this process is a fundamental part of our earthquake design philosophy and it illustrates the difference between our design approach and the traditional design process for low- and high-rise buildings, a few comments about the figure are appropriate.

The conceptual design as discussed in the preceding section is noted as the first phase of the design process in Figure 1.1. The second phase is called the *design development phase*. It is during this phase that the member sizes are selected. Procedures for establishing member sizes must be simple, yet accurate enough to establish spacial requirements and develop the analytical model of the building. Building components are then reinforced to satisfy both the strength and serviceability requirements using the loading criteria established in the conceptual design phase.

The design development phase for most one- to four-story buildings is performed by the engineer by following code criteria. In the existing UBC and ACI code context this means a single-level working stress design approach; in the context of the design procedure proposed here, a code-type working stress design for service load levels and a strength design for extreme load levels. Therefore, a two-level design approach using loads and capacities is developed in a one- to

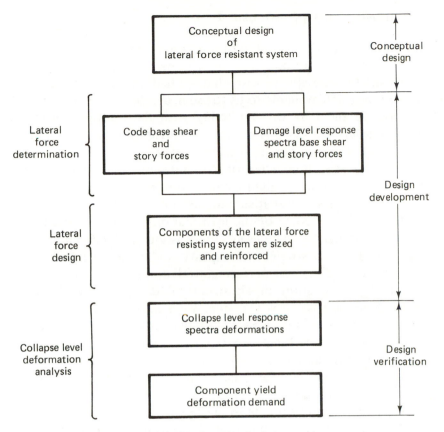

Figure 1.1 Earthquake design procedure

four-story standard for typical buildings. Reference [1.5] presents a draft of such a code developed by the authors and reviewed by a committee of California structural engineers.

The *design verification* phase will be the subject of considerable discussion in Volume 3. This phase is usually beyond the typical scope of services under today's state-of-the-practice design for typical buildings in the one- to four-story height range. Taller buildings or irregularly shaped buildings should be designed using the design verification phase. The advances made in matrix structural dynamics, geotechnical earthquake engineering, and computer software programming enable the design verification phase to answer quite accurately the question: What can happen to the building if this or that extreme loading condition were to occur?

1.4 ORGANIZATION OF VOLUME 2

Structural engineers, by the nature of their profession, must utilize *existing* technical information in order to meet society's needs *today*. This means that we always wish that we had more experimental and analytical research available to us to use in design. Stated differently: We must do the best we can today with the information that is currently available. As this relates to reinforced concrete masonry, let us see what information is available today.

The material properties of concrete masonry have received considerable attention. Significant testing has been performed by private laboratories, universities, and masonry association laboratories. Chapter 2 discusses some of the information available. Specifically, as it relates to earthquake design, there is a considerable amount of data on the ultimate strength of masonry, and modern reinforced concrete masonry designs have performed well during major earthquakes.

At the foundation of all structural design is the theory of structural mechanics. For example, working stress design assumes a linear strain variation across a member cross section, and design equations follow accordingly. Strength design of concrete masonry can be logically developed using existing principles of structural mechanics. Building on the linear strain assumption currently used in working stress design, we can develop equations for member capacities for compression, tension, flexure, and shear. The structural mechanics used is similar, in most respects, to that used in reinforced concrete design. This enables the designer to understand basic assumptions more readily, as well as the associated limitations of the resulting equations. Chapter 3 presents the structural mechanics of concrete masonry.

Analytical models are consciously and unconsciously used to link material behavior and strength (Chapter 2) with structural mechanics (Chapter 3). Simple procedures must be developed to aid in the design of structures that are reliable as well as economical. Chapter 4 discusses assumptions used to develop analytical models commonly encountered in the design of one- to four-story buildings.

With the foundation now set, we can integrate the material in Chapters 2 through 4, and develop a *strength design* approach for concrete masonry. Chapter 5 presents the necessary material to enable the structural engineer to estimate the "nominal," or theoretical, strength of the building members.

We realize that there are many items which must be accounted for in design in addition to the calculation of nominal component strengths. For example, the design material strengths we use in design are unique numbers, and the corresponding values for the as-built members are unique numbers. We also know that the corresponding values for the as-built members are not known prior to construction. Member strengths estimated using the principles of structural mechanics differ from as-built strengths due to the uncertainties in material properties and construction methods as well as a basic uncertainty in the accuracy of the capacity equation. Major advances have recently been made in the methodologies utilized to account for these design uncertainties. Chaper 6 discusses this topic and describes the fundamental considerations that must be taken into account in the development of building code and other structural design criteria. This chapter emphasizes the advantage of using these methodologies to explicitly incorporate current levels of uncertainty in the design process, and explains how the acquisition of additional information can result in reductions in the level of these uncertainties.

Design case studies are presented in Chapter 7. These case studies are intended to demonstrate how the strength design procedure can be used by the structural engineer.

2

MATERIAL PROPERTIES

2.1 GENERAL

Reinforced concrete masonry buildings are built by combining:

1. Concrete masonry units
2. Mortar
3. Grout
4. Steel reinforcement

Each of these elements plays an important role in ensuring that the constructed concrete masonry assembly is capable of resisting anticipated environmental loads.

Figure 2.1 shows a schematic of a concrete masonry wall. The concrete block units are held together by the mortar. Steel reinforcement is placed in the cells of the units, and then the grout bonds the units and reinforcement into a single structural component. This book deals principally with concrete block units; however, the theory and concepts introduced are applicable to other forms of concrete masonry construction.

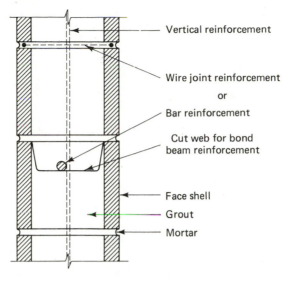

Figure 2.1 Concrete block masonry wall

2.2 CONCRETE BLOCK UNITS

Concrete block units come in a variety of colors, textures, and sizes. The units are usually specified according to the unit width (through the wall thickness) first, unit height next, and unit length last. For example, an 8 × 8 × 16 unit is 8 in. wide, 8 in. high, and 16 in. long. These dimensions are referred to as modular or nominal dimensions and are not actual block sizes. The actual sizes are $\frac{3}{8}$ in. less to allow for the mortar joints. Figure 2.2 shows an 8 × 8 × 16 unit. The standard block is compatible with the special-purpose units which are also shown in the figure. Other common thicknesses are 4, 6, and 12 in., but 10- and 16-in. block may also be obtained by special order.

Concrete block units are constructed of concrete made from a mix of portland cement, sand, lime, aggregate, and water. Units are constructed by placing the mix in molds. The proportions of the mix are varied to control strength. Section 2.6 discusses the methods by which the design strength may be determined.

The designer has several types of concrete masonry units from which to choose. The most common type used in engineered masonry design is hollow load-bearing concrete masonry (ASTM C90). The American Society for Testing and Materials (ASTM) specification

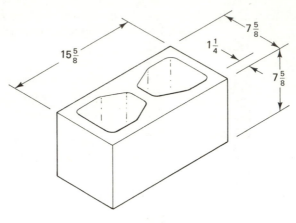

8 × 8 × 16
standard

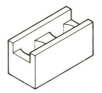

8 × 8 × 16
bond beam

8 × 8 × 8
half

8 × 8 × 16
open end
standard

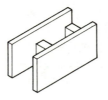

8 × 8 × 16
open end
bond beam

8 × 8 × 8
lintel

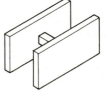

8 × 8 × 16
double open end
bond beam

Figure 2.2 Concrete block unit (8 in. × 8 in. × 16 in.) [2.10]

lists two basic groups consisting of those units intended for exposure
to the elements, above and below grade (Grade N), and units that
may be used only above grade, with a protective coating if exposed
to the weather (Grade S). The Grade N units are most common.

Figure 2.3 shows a stress-strain curve for a concrete masonry specimen obtained from a series of compression tests on concrete masonry walls [2.1] and plotted according to a theoretical curve developed by Kent and Park [2.2]. Examination of the curve shows that the stress is very nearly proportional to the strain for stresses up to at least one-half of the maximum stress. The maximum, or ultimate, compressive stress is labeled f'_m and can be seen as the highest point on the stress-strain curve. Once the maximum stress is reached the stress-strain curve begins a fairly steep descending branch until the ultimate strain, ϵ_{mu}, is reached. The stress at the ultimate strain is seen to be approximately 15 percent below that at f'_m. Referring to an actual stress-strain curve (Figure 2.9) shows that the general shape of the curve shown in Figure 2.3 is a reasonable approximation.

The modulus of elasticity, E_m, of the masonry is constant only over the portion of the curve where the stress is porportional to the strain (i.e., the strain-stress curve is straight). As discussed above, this is essentially true for stresses up to $0.5f'_m$. For this linear region

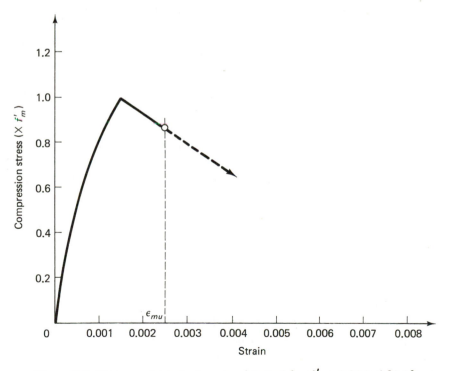

Figure 2.3 Masonry stress-strain curve (plotted for $f'_m = 2.32$ ksi) [2.1]

the value of E_m may be determined from the following expression:

$$E_m = \frac{f_m}{\epsilon_m} \tag{2.1}$$

where f_m = masonry stress at strain ϵ_m

ϵ_m = measured strain

It can be seen from Figure 2.3 that the value of E_m is very nearly $1000 f'_m$.

Current building codes, on the other hand, use an empirical expression to determine the modulus of elasticity of concrete masonry. These expressions do not explicitly consider the type of concrete block or mortar used in the structure. The *Uniform Building Code* [2.3] assumes that E_m is a function of the ultimate compressive stress (f'_m) of the total masonry, mortar, and grout system, and recommends the following equation:

$$E_m = 1000 f'_m \leqslant 2{,}000{,}000 \text{ psi} \tag{2.2}$$

There is currently some discussion regarding the modification of Equation (2.2) to set the value of E_m equal to 600 times the compressive strength of the masonry. Recent tests on concrete masonry wall specimens suggest that the actual value of E_m may actually be closer to $1000 f'_m$ for stresses of at least one-half of f'_m [2.1] and that this proposed reduction is overly conservative. It should be remembered that concrete masonry is an intermittently loaded material, and the long-term effects of sustained load are quite small.

In a similar manner, an empirical expression is used to determine the shear modulus of concrete masonry (E_v). The value of E_v is assumed equal to $0.4 E_m$. This is the same ratio used to set the shear modulus for concrete; thus,

$$E_v = 400 f'_m \tag{2.3}$$

2.3 MORTAR

Over 4000 years ago the Great Pyramids of Giza in Egypt were constructed using masonry mortars. The massive stone blocks forming the pyramids were cemented together with a mortar made of burned gypsum and sand. In ancient Greece and Italy, mortars were produced from various materials, such as burned lime, volcanic tuff, and sand. Although many centuries passed before mortar was used in the first settlements in North America, the product had not developed beyond the point where a relatively weak mortar was being made from lime and sand. Strengths of mortar increased significantly when portland cement was used to make mortar in the early part of the twentieth century. More recently, a product called *masonry cement* was developed to make mortars with excellent strength characteristics.

The relatively small proportion of mortar in concrete masonry construction significantly influences the total structural performance. The mortar binds the individual masonry units together and seals the joints against penetration by air and moisture.

Portland cement, a hydraulic cement, is the principal cementious ingredient in masonry mortars. Three types are permitted by ASTM specifications:

Type I: for general-purpose construction

Type II: for moderate resistance to sulfate action

Type III: for high-early-strength development

Masonry cement is a mixture of portland cement and finely ground inert materials such as hydrated lime in approximately equal amounts. Other agents may be added to influence properties such as plasticity, water retentivity, and durability.

Aggregates for mortar may consist of natural or manufactured sand. Manufactured sand is obtained by crushing stone, gravel, or air-cooled blast furnace slag. It is characterized by sharp, angular particles. Well-graded aggregate reduces separation of the components in a plastic mortar mix, and in turn, this reduces bleeding and improves workability.

Water for masonry mortar is required to be clean and free of deleterious amounts of acids, alkalis, or organic materials. Whether the water is potable is not, in itself, a consideration, but the water obtained from drinking supply sources is generally suitable for mixing mortar.

Each of the principal constituents of mortar (portland cement, lime, sand, and water) makes a definite contribution to the performance of the mortar. Portland cement contributes to strength and durability. Lime, which sets only on contact with air, gives the mortar workability, water retentivity, and elasticity. Both lime and portland cement contribute to bond strength. Sand acts as a filler which also contributes to the strength of the mix. It greatly decreases the setting time and drying shrinkage of mortar, thereby reducing cracking. The presence of sand enables the unset mortar to retain its shape and thickness under several courses of concrete masonry units. Water is the mixing agent that gives workability and hydrates the cement.

Mortar produced from masonry cement or a combination of portland cement and lime is given an ASTM mortar designation based on the proportioning of materials as prescribed in Table 2.1. The aggregate, measured in damp, loose condition, should not be less than $2\frac{1}{4}$ and not more than 3 times the sum of the volumes of the cement and the lime used. A combination of portland cement and lime will suffice as the cementing agent in each type of mortar listed. Masonry cement alone is suitable only for Types O and N mortar. Masonry cement can be used in mortar Types M and S only when it is combined with portland cement because a higher portland cement content is required to produce high-strength mortars.

Mortar Type M is specified when the masonry will be subjected to high compressive loads, severe frost action, or high lateral loads from seismic, hurricane, or earth loads. It is also used in structures constructed below grade. In applications requiring normal compressive strength but high flexural bond strength, Type S is specified, and is the most common type of masonry mortar. The building codes do not explicitly recognize the added strength of the higher-strength mortars except in unreinforced masonry design; however, the UBC

TABLE 2.1 Mortar Proportions by Volume

Mortar Type	Parts of Portland Cement by Volume	Parts of Masonry Cement by Volume	Parts of Lime by Volume	Parts of Aggregate by Volume
M	1	1 (Type II)	0	5.0–6.0
	1	0	$\frac{1}{4}$	2.8–3.8
S	$\frac{1}{2}$	1 (Type II)	0	3.4–4.5
	1	0	Over $\frac{1}{4}$ to $\frac{1}{2}$	2.8–4.5
N	0	1 (Type II)	0	2.3–3.0
	1	0	Over $\frac{1}{2}$ to $1\frac{1}{4}$	3.4–6.8
O	0	1 (Type I or II)	0	2.3–3.0
	1	0	Over $\frac{1}{4}$ to $1\frac{1}{2}$	2.8–7.5

TABLE 2.2   Mortar Selection Chart [2.3]

Mortar Type	Construction Applications	Permitted by the *Uniform Building Code* in Reinforced Concrete Masonry Construction
M	High compressive loads, high lateral loads, severe frost action, below-grade applications	Yes
S	Normal compressive loads, high lateral loads; most common type in reinforced masonry	Yes
N	Residential construction, masonry veneers, interior construction	No
O	Non-load-bearing walls and partitions	No

permits the use of only Type M and S mortars in reinforced concrete masonry.

Residential construction with masonry veneers applied to a wood frame, basement, and interior walls typically uses Type N mortar. Type O is specified for construction with non-load-bearing walls and partitions.   Table 2.2 summarizes the construction applications of each type of mortar.

The strength of masonry construction usually increases with the strength of the mortar, but mortar strength achieved beyond that required for structural adequacy of the masonry generally contributes little to the strength capacity of the masonry assembly.   Table 2.3 gives representative values of the compressive strength of test cubes for the five different mortar types.

Workability is the property of mortar characterized by a smooth, plastic consistency which makes it easy to spread. Workable mortar holds the weight of concrete blocks when placed and makes alignment easy.   It adheres to vertical masonry surfaces and readily squeezes out of mortar joints.   Water affects the workability of a mix by influenc-

TABLE 2.3   Average Compressive Strength of Lab Mortar Cubes Made per ASTM C270

Mortar Type	Average Compressive Strength of Cubes (psi)	UBC Minimum Field Compressive Strength (psi)
M	2500	1500
S	1800	1500
N	750	—
O	350	—

ing the consistency. A well-graded, smooth aggregate improves work-ability as previously discussed. Air entrainment adds to workability through the action of the minute air bubbles, which function like ball bearings in the mixture. An increase in lime content also im-proves workability by increasing the capacity for water retentivity. Water retentivity is a measure of the ability of a mortar to retain its plasticity when in contact with an absorbent unit. This gives the mason time to place and adjust the block before the mortar stiffens.

Properties of hardened mortars which affect the performance of the finished concrete masonry include compressive strength, bond strength, and durability. If one compares the compressive strength of the mortars shown in Table 2.3 with the proportions shown in Table 2.1, it can be seen that the compressive strength increases as the cement content is increased while it decreases as the lime content is increased. The UBC requires the minimum field compressive strength of Types M and S mortar to be 1500 psi. Air-entraining agents, added to the mortar to increase its workability and improve its resistance to frost action, can also reduce the compressive strength of the mortar if excessive amounts are added.

With hardened mortar, compressive strength is the most easily determined property. Compressive strength tests as prescribed in ASTM C270 are made on 2-in. cubes cast in nonabsorbent molds and cured in water or moist air. Field-prepared mortar exhibits different characteristics than laboratory mixed mortar principally because of the absorptive properties of masonry units. When mortar is placed on a masonry unit some of the water in the mortar is rapidly removed by the unit, which reduces the water-to-cement ratio. A decrease in the water/cement ratio causes an increase in the compressive strength of the mortar, as seen in Figure 2.4. The suction properties of the masonry unit, as well as temperature and humidity conditions, influ-ence the rate of moisture loss. Dry weather increases the rate of moisture loss, and damp weather prolongs the period of hydration. If the concrete masonry has a very low moisture content, the water may be removed from the mortar so quickly that a very poor bond results.

With concrete masonry construction, the compressive strength of mortar has little influence on the compressive strength of a wall or assemblage. In one series of tests, a decrease of 69 percent in the compressive strength of the mortar resulted in a decrease in the prism strength of less than 10 percent [2.4].

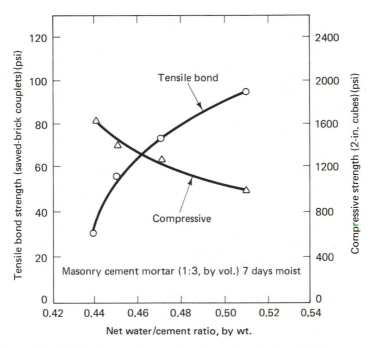

Figure 2.4 Effect of water/cement ratio on tensile bond and compressive strengths

Bond is the property of hardened mortar that holds the masonry units together. Bond strength is possibly the most significant single property of hardened mortar. It also is the most capricious and unpredictable. Since bond strength is much smaller than the compressive strength, mortar joints subjected to relatively small tensile stresses are likely to fail. For example, mortar having a laboratory test compressive strength of 2500 psi may develop a bond strength of only 50 to 100 psi. A lack of bond at the interface of mortar and concrete block may lead to moisture penetration through the unbonded areas. Other factors being equal, the bond strength of mortar will increase as compressive strength increases. A complete and intimate contact between mortar and block surface is essential for good bond and can be achieved through use of mortar having good workability.

Workmanship is paramount in effecting bond strength. The time lapse between spreading mortar and placing block should be kept to a minimum because the flow will be reduced through suction of the block on which it is first placed.

Figure 2.4 shows that an increase in the water/cement ratio causes an increase in the tensile bond strength. Optimum bond strength is obtained with mortar having the highest water content compatible with workability. The moisture content of the concrete masonry also influences the tensile bond strength. If the masonry has a very low moisture content, the water may be removed from the mortar so quickly that complete hydration cannot take place and a poor bond will result. Rapid removal of the water from the mortar also decreases the workability of the mortar.

Many substances have been employed as admixtures, such as tallows and salts of wood resins, and various chemical admixtures for masonry mortars are available. Since admixtures are commercially prepared products, their compositions are not generally disclosed in full. Few data have been published regarding the effect of propietary admixtures on mortar bond or strength, but field experience indicates that detrimental results have occurred in some cases. For this reason, admixtures of unidentified composition should be used only after it has been established by test or experience that they do not materially impair the mortar performance.

Admixtures are added to mortar to improve certain characteristics. In very hot, dry weather, an admixture may be added to retard the setting time to allow the masons a little more time before the mortar must be retempered. On the other hand, admixtures are added to mortar in cold climates to accelerate the hydration of the cement in the mortar.

There are several admixtures that should not be used in portland cement mortars. Calcium chloride is sometimes added to accelerate the setting of mortar; however, this admixture often corrodes the steel rebar in reinforced masonry. Gypsum has also been used to speed up the set, but this type of mortar has been known to undergo premature disintegration.

2.4 GROUT

Grout is a mixture of cementitious material, aggregate, and enough water to cause the mixture to flow readily, without segregation, into cores or cavities in the concrete masonry units.

In reinforced concrete masonry wall construction, grout often is placed only in wall spaces containing steel reinforcement. The grout bonds the masonry units and steel so that they act together to resist imposed loads and it must be placed in cells containing reinforcing steel. In some reinforced load-bearing masonry walls, all cores, those with and without reinforcement, are grouted to further increase the strength of the wall.

Sometimes grout is used to give added strength to unreinforced load-bearing concrete masonry walls. This is accomplished by filling some or all of the cores with grout (e.g., filling the cores of the units forming the jamb of a door). It is also used to fill bond beam cavities and occasionally to fill the collar joint of a two-wythe (i.e., two units thick) wall.

Grout for use in concrete masonry walls should comply with ASTM C476, "Specifications for Mortar and Grout for Reinforced Masonry." Table 2.4 lists the proportions of the ingredients in the grout which meet the requirements of this specification.

The choice of fine or coarse aggregate in the grout depends mainly on the minimum horizontal cross-sectional dimensions of the grout space. Building codes vary with respect to minimum dimensions of grout spaces and maximum size of aggregate in grout; however, the smallest space to be grouted should be at least 2 × 3 in. In this case, fine grout (sand aggregate) should be used. When the minimum horizontal dimension of the space is at least 4 in., use a coarse grout containing pea gravel. Pea gravel, with a maximum aggregate size of $\frac{3}{8}$ in., produces a grout that is strong and less expensive than one using fine grout. Coarse aggregate grout should also be used for two-wythe masonry walls if the continuous clear space between wythes is at least 2 in.

If the minimum dimensions of the grout space exceed 5 to 6 in.

TABLE 2.4 Grout Proportions by Volume

Type	Parts by Volume of Portland Cement	Parts by Volume of Lime	Aggregate, Measured in a Damp, Loose Condition	
			Fine	Coarse
Fine aggregate grout	1	0 to $\frac{1}{10}$	$2\frac{1}{4}$ to 3 times sum of volumes of cement and lime	None
Coarse aggregate grout	1	0 to $\frac{1}{10}$	$2\frac{1}{4}$ to 3 times sum of volumes of cement and lime	1 to 2 times sum of volumes of cement and lime

(e.g., in pilasters or columns) the cavities can be filled with conventional concrete with a maximum aggregate size of 1 in. Some specifying agencies stipulate that $\frac{3}{4}$-in. gravel should be used when the grout space is greater than 4 in.; however, the choice of the maximum size of the aggregate should be consistent with the particular job conditions to ensure satisfactory placement of the concrete fill and proper embedment of the reinforcement in the grout.

All grout should be of fluid consistency; that is, it should be as fluid as possible without segregation of the materials. When the slump is measured using ASTM C143, "Method of Test for Slump of Portland Cement Concrete," the desired range is from 8 in. up to 10 in.

Wherever possible, grout or concrete fill should be batched, mixed, and delivered in accordance with the requirements for transit-mixed concrete, ASTM C94, "Specification for Ready-Mixed Concrete." Transit-mix grout should be rotated continuously from the time the water is added until placement. When a batch mixer is used on the job site, all materials should be mixed thoroughly for at least 5 minutes. Grout not placed within $1\frac{1}{2}$ hours after the water is first added should be discarded.

The ASTM specification for grout (ASTM C476) does not provide for field control because it is a proportion specification; however, some building codes require that compressive strength tests of grout be made in the field. When this test is required, the following procedure is recommended for making the specimens.

First, select a flat location where the block mold will remain undisturbed for 48 hours. Place a piece of wood $1\frac{5}{8}$ in. thick and 3 in. square on the level surface. Four masonry units with permeable paper, such as absorptive paper toweling, taped to one face shell are placed around the wood block to form the mold. The absorbtive paper prevents the grout from bonding to the masonry and allows the excess moisture to be absorbed by the block to reduce the water/cement ratio. This reduction in the water/cement ratio produces test specimens which more accurately model the in-situ grout. The resulting mold is approximately 3 × 3 × 6 in. high (Figure 2.5).

The number of grout samples to be taken should be specified before the start of construction. If the number is not specified, take one sample and cast two specimens for each 30 yd^3 of grout or fraction thereof being placed each day. Also take a sample whenever there is any change in mix proportions, method of mixing, or materials used.

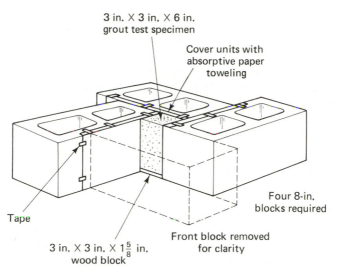

3 in. × 3 in. × 6 in.
grout test specimen

Cover units with
absorptive paper
toweling

Four 8-in.
blocks required

Tape

3 in. × 3 in. × 1$\frac{5}{8}$ in.
wood block

Front block removed
for clarity

Figure 2.5 Grout specimen [2.12]

Methods for grouting the wall will vary with the job size, available equipment, and the masonry contractor's preference and, in turn, influence the way in which the masonry units are laid up and the reinforcing steel is placed.

There are two grout placement procedures in general use: (1) low-lift grouting, where the grout is placed in lifts or pours up to 4 ft in height and no cleanout holes are needed; and (2) high-lift grouting, where grout is placed in lifts of story height or higher and cleanout holes are required at the bottom of each grout space containing reinforcement.

A *lift* is the layer of grout placed in a single continuous operation. A *pour* is the entire height of grout placed in 1 day and may be composed of a number of successively placed grout lifts. When all of the grout is to be placed at one time, it is essential that equipment capable of delivering the required volume be available. This requirement can often be met economically through the use of a grout pump, which facilitates delivery of the grout to the desired location with a minimum of segregation in the grout.

Low-lift grouting is the simplest method of grouting concrete masonry and is more common on smaller jobs. Low-lift grouting requires no special concrete block shapes or equipment. The wall is built to scaffold height, or to a bond-beam course, and steel reinforcing bars are then placed in the designated vertical and horizontal cells of the concrete masonry. The cells are then grouted, with the level of the grout being stopped at least $1\frac{1}{2}$ in. from the top of the masonry. The steel reinforcing should project above the top course a minimum of 30 bar diameters to ensure a proper lap splice with the steel added for the next grout lift.

In high-lift grouting, the grouting operation is postponed until the concrete masonry wall is laid up to full story height. This method allows the mason the option of placing the vertical wall steel before or after the masonry block is all in place. If the steel is placed first, the mason will usually use an open-ended block to simplify the threading of the block around the steel. Placement of the vertical steel after the block has been laid eliminates the need for the open-ended block, but requires positioners to keep both the vertical and horizontal reinforcing steel in the desired location. If the vertical steel is placed before the wall is laid up, positioners are not required because the vertical steel is tied to the horizontal steel to assure proper alignment.

Both high-lift and low-lift grout require that the grout bond to the masonry units and to the steel. Mortar droppings on these elements can prevent a good bond and compromise the integrity of the member. To prevent this from happening, cleanouts are provided at the base of the walls to facilitate the removal of this debris prior to grouting. A special scored unit is sometimes used to permit easy removal of part of the face shell for cleanout openings.

A major concern in high-lift grouting is the prevention of *blowouts*. Blowouts occur when the hydrostatic pressure of the wet grout causes the wall to deflect or collapse laterally. Blowouts causing deflections of even $\frac{1}{8}$ in. out-of-plumb require that the wall be torn down and rebuilt because the bed joint has been destroyed by the lateral movement, and pushing the wall back into plumb will not rebond the units. Blowouts can be prevented by adequately bracing the walls and by grouting in fairly small lifts. Common practice suggests that the grout be placed in 4-ft lifts with a 15- to 60-minute waiting period between lifts. This waiting period allows some of the excess moisture to be absorbed into the concrete block units and reduces the hydrostatic pressure.

Vibration of the grout is another important factor in assuring a competent job in both high- and low-lift grouting. Just as loose mortar droppings can prevent the formation of the grout bond, so, too, can improper vibration. The excess water added to the grout which allows it to flow so readily is usually absorbed quite quickly by the concrete block units. The removal of the extra moisture from the grout causes it to shrink and to pull away from the reinforcing. Reconsolidating the grout lift between pours brings the grout back into contact with the steel and assures a good bond.

Vibration is also important in grouting for many of the same reasons it is essential in cast-in-place concrete. Voids that are formed in the grout can be removed by proper vibration. Vibration usually is carried out by mechanical vibrators, but strict control is necessary to prevent over vibrating the grout and causing it to segregate. One-by two-inch wooden sticks are also frequently used to consolidate the grout.

Current design methods do not reflect the strength of grout above the code minimum. The UBC simply requires that grout attain a minimum compressive strength of 2000 psi in 28 days. This neglect of the compressive strength of the grout does not imply that grouted masonry construction is not a reasonable construction system. On the contrary, reinforced concrete block masonry would not be possible without grout. Often, the compressive stress block in a concrete masonry wall in flexure occurs in the face shell of the unit, not in the grout space. Consequently, the grout is usually in tension and serves to transfer the stress to the steel. Current design practice does not usually involve determining if the compressive stress block includes a portion of the grout space to ascertain if higher-strength grout could provide an increase in section capacity.

2.5 STEEL REINFORCEMENT

The development of reinforced concrete masonry has a parallel in the development of reinforced concrete. Both systems are heterogeneous, that is, made up of more than one material, each of which has different properties. Concrete masonry, like concrete, is excellent in resisting compressive forces but is relatively weak in tension. Reinforcing steel, on the other hand, is subject to buckling problems when under compression but performs excellently when used to resist tension forces. The combination of these two materials, concrete masonry for compression and steel for tension, produces a homogeneous structure capable of resisting applied loads.

Reinforced concrete masonry performs as a homogeneous section because of the strain compatability between the steel, grout, and concrete masonry units. This strain compatability is assured by the bonding of these materials. Bond is produced by the naturally rough surface and ribbed deformations on the steel bars and the chemical adhesion between the surface of the steel and the surrounding grout, which in turn bonds to the concrete masonry units. The similarity in the coefficients of thermal expansion of steel, grout, and concrete masonry reduces the buildup of disruptive thermal stresses at the interface between the materials, which could lead to bond failure.

The selection of reinforcing steel for a masonry member is based on several considerations, two of which are available sizes and strengths. Currently, reinforcing steel is produced in deformed bars ranging from No. 3 ($\frac{3}{8}$ in. in diameter), to a recommended maximum of No. 11 bars (approximately $1\frac{3}{8}$ in. in diameter). Reinforcing steel may be either Grade 40, with a minimum yield strength of 40,000 psi, or Grade 60, with a minimum yield strength of 60,000 psi.

The selection of steel involves more than considering what is available. If an allowable stress method (working stress design) is used in concrete masonry design, the compressive capacity of the masonry is the governing factor in most cases. The use of a high-strength steel (Grade 60) is not really justified because little of its additional strength is recognized in the working stress method. On the other hand, if the designer elects to use a strength design method, the tensile capacity of the steel is a prime consideration in proportioning the member, and the use of a high-strength steel might then be advisable.

The designer should also consider the realities of the construction sequence. If certain bars must be bent (such as dowels), a ductile steel is recommended. Grade 40 steel possesses much greater ductility than does Grade 60 and can withstand abuse without snapping off.

The maximum bar sizes limited by building codes will also influence the designer's choice of steel. Again, considering the realities of construction, building codes try to prevent the designer from using bars of such a large diameter that the flow of grout around the bar is impeded and voids form, reducing the member's strength.

ASTM specifications cover the physical characteristics of the reinforcing steel, such as minimum yield stress, and require identification marks to be rolled into the surface of one side of the bar to denote the producer's mill designation, bar size, type of steel, and grade marks indicating yield strength. Figure 2.6 shows examples

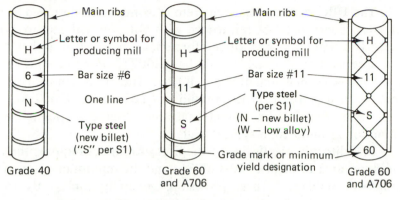

Steel Type
N for new billet
S for supplemental requirements A615
A for axle
1 for rail
W for low alloy

Figure 2.6 Reinforcing steel marking system [2.13]

of identification marks for Grade 40 and Grade 60 steels. Grade 40 bars show only three marks (no grade mark), in the following order:

1. Producing mill (usually an initial)

2. Bar size number (Nos. 3 through 11)

3. Type N for "new billet," A for "axle," I for "rail," W for "low alloy" (Type N is most common)

Grade 60 bars must also show grade marks:

60 (or One Line) for 60,000 psi strength

Grade mark lines are smaller and between the two main longitudinal ribs, which are on opposite sides of all U.S.-made bars. Number grade marks are fourth in order.

High-strength steel wire fabricated in ladder or truss systems is often placed in the bed joints to reinforce the wall in the horizontal direction. The most common uses of joint reinforcing are to control shrinkage cracking in concrete masonry walls. The joint reinforcing is part of the minimum steel reinforcing required by the building codes. It can also be used as a continuous tie system for faced, veneered, and cavity walls. Although the UBC does not explicitly recognize the structural contribution of joint reinforcing, it is still present and adds to the lateral load-carrying ability of the wall by tieing it together.

The UBC requires that concrete masonry walls designed as reinforced masonry must be reinforced with both vertical and horizontal steel. The minimum area of reinforcement in a wall in either direction, vertical or horizontal, should be 0.0007 (0.07 percent) times the gross cross-sectional area of the wall (bt). The minimum total area of steel, vertical and horizontal, should not be less than 0.002 (0.2 percent) times the gross cross-sectional area. Figure 2.7 identifies the dimensions b and t for a typical wall.

In California, the Office of the State Architect requires the total minimum area of steel to be $0.003bt$ and the minimum in one direction to be $0.001bt$. These provisions generally apply only to masonry construction used in schools and hospitals.

The minimum area of wall steel is an arbitrary amount that has a parallel in the minimum steel requirement for reinforced concrete walls. For example, the minimum area of steel required for a concrete

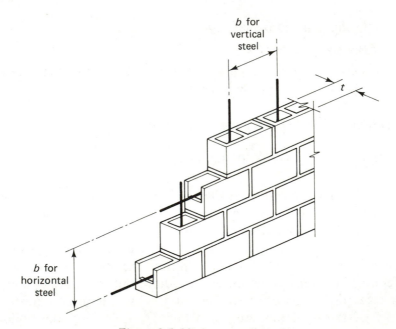

Figure 2.7 Minimum wall steel

wall is usually $0.0025bt$ in the horizontal direction and $0.0015bt$ in the vertical direction, for a total of $0.004bt$. This is twice the amount required for reinforced concrete masonry; however, concrete is cast in a plastic condition and is subject to significant shrinkage during hydration. Concrete masonry units, on the other hand, have already assumed their final shape and size and only mortar and grout are added to the wall section. Because there is far less material to shrink in a concrete masonry wall compared to a concrete wall of equivalent thickness, the requirement for minimum steel was set at half that of concrete.

Although there may be less material to shrink in a concrete block wall, shrinkage considerations will probably govern the orientation of the maximum amount of required minimum steel. Shrinkage is much greater in the horizontal direction in grouted concrete masonry than in the vertical direction, and because of the transfer of water from the grout cell walls, the grout attempts to pull away from the block. This shrinkage either leads to a bond failure or a vertical crack, neither of which is desirable. Horizontal steel can relieve these shrinkage-induced problems by evenly distributing the stress throughout the entire length of the wall. Consequently, unless a particular design dictates otherwise, the maximum amount of required minimum steel should probably be distributed as horizontal steel.

The prevention of cracking in the masonry from shrinkage or thermal stresses can also be accomplished through the use of crack control joints. A control joint is a weakened plane for the full height of the wall. This is usually constructed by using a $\frac{3}{8}$-in.-wide $\times$ $\frac{3}{4}$-in.-deep joint up the height of the wall and filling the resulting recess with an elastomeric caulking compound. The horizontal steel must be terminated at the joint; however, the reinforcing steel at the ledger (typically at a roof or floor line) must remain continuous. This provision is required because a crack control joint creates a weakened plane in the masonry and may interfere with the seismic performance of the wall.

There are several common locations for control joints:

1. At major changes in wall height
2. At changes in wall thickness (other than at pilasters)
3. At control joints in the roof, floor, and foundation
4. At openings through the wall

5. At sections of reduced cross section where mechanical chases or structural columns are installed

Several other factors influence the location and spacing of the control joints. The length of the wall and the distribution of seismic forces to the wall are two important additional factors. In a typical wall, the control joints are generally spaced from 15 to 50 ft on center, with a 20-ft spacing being common. Table 2.5 suggests some control joint spacing for Type I concrete masonry units.

Another reason for requiring minimum steel reinforcement is to ensure ductile behavior of the wall in the event of excessive overloads due to extreme lateral loads such as earthquake forces. The minimum steel that must be provided in one direction is $0.0007bt$. Calculations show that a wall reinforced in this manner has a capacity of more than 80 psi in axial tension due to flexure. The UBC allows 25 psi tension for inspected unreinforced grouted concrete masonry, plus a one-third increase for earthquake forces, for a maximum of 33 psi. Even though the wall must crack to stress the steel, a wall reinforced with the minimum amount of steel is nearly 2.5 times stronger than an unreinforced wall (80 psi versus 33 psi). Ductile behavior of the wall is assured because the steel will yield long before the concrete masonry reaches its full compressive strength.

Although the ultimate load-carrying capacity of unreinforced concrete masonry walls is significant, they have little or no reserve strength and are subject to sudden collapse if the ultimate load capacity is exceeded. Reinforcing steel not only increases the ultimate load capacity of the wall, but also greatly improves its inelastic behavior and thereby reduces the probability of a sudden collapse.

Unlike the minimum steel requirement for walls, the UBC does not specify a minimum steel ratio for flexural elements. The steel ratio is the area of the steel parallel to the plane of bending divided

TABLE 2.5   Control Joint Spacing for Type I Concrete Masonry Units

	Vertical Spacing of Joint or Horizontal Reinforcement (in.)			
	48	24	16	4
Recommended control joint spacing expressed in terms of panel height-to-length ratio, H/L	2 ($L \leqslant 40$ ft)	$2\frac{1}{2}$ ($L \leqslant 45$ ft)	3 ($L \leqslant 50$ ft)	4 ($L \leqslant 60$ ft)

by the area of the flexural masonry. The minimum amount of flexural steel in reinforced concrete members required by the building code is a function of the strength of both the concrete and the steel. These requirements for concrete are intended to permit the concrete member to have tensile strength in the steel which exceeds the cracking moment of the concrete section. A similar well-defined philosophy does not presently exist in the design of reinforced concrete masonry. Prudent design practice might suggest that the designer determine the cracking moment of a flexural member and provide tensile reinforcement sufficient to produce a tension force which exceeds this value but is less than the compressive strength of the concrete masonry.

Reinforcing steel in columns contributes to the load-carrying capacity of the member because the horizontal ties around the vertical steel inhibit buckling of the reinforcing steel. Vertical wall steel is assumed not to take a vertical load because it is not restrained against buckling by ties; however, vertical wall steel does participate in load-carrying capacity and is an uncounted added factor of safety in a reinforced concrete masonry wall. The minimum steel ratio for columns should not be less than 0.005 (0.5 percent) and there should not be fewer than four No. 3 bars in a column. The maximum steel ratio should be no more than 0.04 (4 percent), and the maximum-size bars should be No. 10. In this case, the steel ratio is the area of steel divided by the gross area of column.

Although the specified yield stress for Grades 40 and 60 reinforcing steel is 40 ksi and 60 ksi, respectively, the actual yield stress will typically exceed these values. Research shows that a reasonable mean value of the yield stress, f_y, for Grade 60 steel is 71.5 ksi with a coefficient of variation of 7.7 percent. Similarly, the mean yield stress of Grade 40 steel is 48.7 ksi and the coefficient of variation is 6.6 percent [2.5]. Chapter 6 discusses how this information may be used to quantify the uncertainty present in the design of a reinforced concrete masonry member to estimate its actual ultimate strength and establish its structural reliability.

2.6 CONCRETE MASONRY PRISMS

In order to use "engineered" concrete masonry design criteria for structural design of load-bearing concrete masonry walls, it is necessary to know the ultimate strength of concrete masonry or its strength at the earliest age at which it is expected to carry full loading. The ultimate compressive strength of concrete masonry, f'_m, is the basis for allowable axial and flexural compressive stresses in engineered masonry design and is a function of the ultimate compressive strength of the individual concrete masonry units, mortar, and grout. Design codes show coefficients to be used with f'_m to arrive at allowable stresses.

Three methods are used to determine the ultimate compressive strength of the masonry:

1. Compressive load testing of full-scale walls, usually 4 ft (1.2 m) wide by 8 ft (2.4 m) high, conducted in accordance with "Methods of Conducting Strength Tests of Panels for Building Construction," ASTM E72. (This is basically a research method. Because it is time-consuming and expensive, it has been replaced by other methods for use in design.)

2. A long-accepted empirical method which assumes a value for masonry strength based on the average strength of the concrete masonry units.

3. Prism tests made in advance of the design using materials and workmanship similar to those to be used in a particular structure.

As shown in Table 2.6, the designer may use an assumed value of f'_m based on the compressive strength of individual units to be

TABLE 2.6   Assumed f'_m Values for Concrete Block Units
(No Prism Test) [2.6]

Compressive Test Strength of Masonry Units (psi) on the Net Cross-Sectional Area	Compressive Strength of Masonry,[a] f'_m (psi)	
	Type M and S Mortar	Type N Mortar
6000 or more	2400	1350
4000	2000	1250
2500	1550	1100
2000	1350	1000
1500	1150	875
1000	900	700

[a]Values of f'_m may be interpolated but not extrapolated.

used without conducting prism tests. The compressive strength of the individual concrete masonry units is greater than that of a wall built with the same units. Assumed compressive strength of concrete masonry based on unit strength is conservative, especially in the higher-strength categories.

For quality control during construction, three concrete masonry units should be selected from each lot of 10,000 units or fraction thereof and tested in compression in accordance with "Methods of Sampling and Testing Concrete Masonry Units," ASTM C140. For lots of more than 10,000 but fewer than 100,000 units, six units should be tested in compression. When the lot contains more than 100,000 units, three units should be tested for each 50,000 units or fraction thereof in the lot.

A prism is a small assembly of two or more concrete masonry units mortared on top of one another, usually in a stack bond pattern (Figure 2.8). Tests of prisms may be used to determine the ultimate compressive strength of the concrete masonry, f'_m, for single- or multiwythe walls and composite, faced, or cavity walls. The compressive strength of concrete masonry determined from the prism test generally is higher than that based on an assumed value of f'_m; however, the compressive strength of the prism is less than the compressive strength of individual units because of the presence of mortar joints and slenderness ratio effects. Experience has shown that prism strength for hollow units is from 70 to 80 percent of the unit strength when only face-shell mortar bedding is used, and from 80 to 90 percent when full mortar bedding is used.

In 1972 the ASTM E447 specification, covering test methods for compressive strength of masonry prisms, was issued and describes two methods for testing prisms:

Method A: for determining comparative data on the compressive strength of concrete masonry built in the laboratory with either different concrete masonry units or mortar types, or both

Figure 2.8 Typical concrete block prism

Ungrouted, stack bond prism: two courses

Method B: for determining the compressive strength of concrete masonry built at the job site with the same materials and workmanship to be used, or being used, in a particular structure

ASTM E447 also covers testing of concrete masonry units, mortar, and grout used in prisms.

Making and testing prisms during the course of construction is very similar to the quality control measures used in concrete construction. The prisms provide the same type of data as concrete test cylinders made during a job by ensuring that specified strengths are being met.

Prior to wall construction, preliminary tests to establish the design stresses should consist of not fewer than three prisms. When the compressive strength of concrete masonry prisms is used for quality control or inspection at the site, a sample should be tested consisting of three specimens for each 5000 ft^2 of wall area. At least three test prisms should be tested for any building.

Test prisms of concrete masonry units usually have a nominal height-to-thickness ratio, h/t, of not less than 2 and are a minimum of 15 in high. The thickness of a prism must be the same as the thickness of the wall in the structure. The length of the prism must equal or exceed its thickness. For example, if the wall thickness is 8 in., the prisms may be built in the form of squares 8 in. by 8 in. in plan and 16 in. high or in the form of rectangles 8 in. by 16 in. in plan and 16 in. high. The preferable length of the prism is one unit. If the h/t ratio differs from 2, multiply the compressive strength of the specimens determined by test, f'_m, by the appropriate correction factor shown in Table 2.7.

The compressive strength, f'_m, of each specimen should be calculated by dividing the ultimate test load by the cross-sectional area of the prisms, excluding all unfilled cores or voids. The value of f'_m used for design should be the average of the three specimens tested but should not be more than 125 percent of the minimum value determined by the tests. Figure 2.9 shows a typical stress-strain curve for a concrete block masonry prism.

TABLE 2.7   Masonry Prism Correction Factors [2.3]

Ratio of Height to Thickness, h/t	1.5	2.0	2.5	3.0
Correction Factor[a]	0.86	1.00	1.11	1.20

[a]Interpolate to obtain intermediate values.

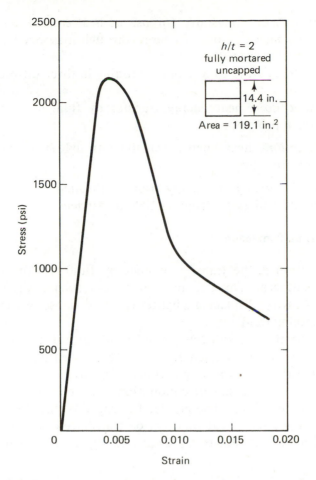

Figure 2.9 Typical stress-strain curve for concrete masonry prisms [2.7]

The compressive strength of all job-site quality control prisms should exceed 0.80 of the value of f'_m used for design and the average compressive strength of all the tests should exceed f'_m.

2.7 ENERGY TRANSMISSION CHARACTERISTICS OF CONCRETE MASONRY

Although the structural adequacy of concrete masonry is an extremely important consideration in the design of a building, increasing national concern over the use and cost of energy suggests that energy conservation is another aspect in design. In this section the mechanics of heat transfer through concrete masonry walls, the thermal advantage of masonry for use as heat storage, and current methods to impede the loss of energy through masonry walls to the outside are briefly discussed.

Heat is transferred to or from a building in three different ways:

1. *Convection:* heat energy transfer by fluid motion, as in a mixing process
2. *Radiation:* heat energy transfer through space by electromagnetic waves
3. *Conduction:* heat energy transfer through materials (solids, liquids, and gases) from particle to particle

Heat Transfer by Convection

Convection is the transfer of heat by fluid motion in contact with solid surfaces. This occurs when a fluid, such as air, is put into motion by gravity, displacing a lighter (i.e., less dense) warm fluid by a heavier cooling fluid.

Convection can also occur as the result of a difference in pressure caused by mechanical energy (e.g., forced ventilation) or large-scale barometric variations (e.g., wind). The transfer of heat energy actually occurs as a result of conduction between the fluid and the surface of the material. The greater the speed of the fluid flow, the greater the amount of heat that can be transferred in a given period of time. Figure 2.10 illustrates several examples of heat transfer by convection.

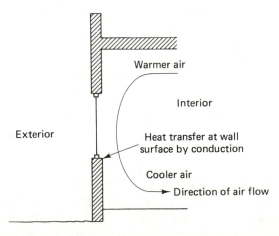

Figure 2.10 Heat transfer by convection caused by interior airflow

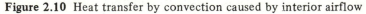

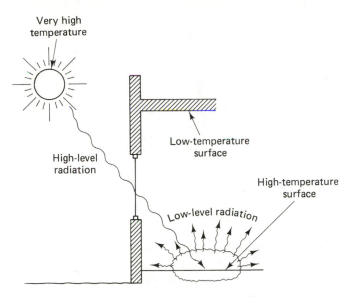

Figure 2.11 Heat transfer by radiation

Heat Transfer by Radiation

Radiation involves the transfer of heat energy between two separate bodies by electromagnetic waves. Unlike conduction and convection, which require a solid medium to accomplish the transfer of energy, radiant heat is transferred from the body at a higher temperature to the body at the lower temperature regardless of whether the space between the bodies contains an atmosphere or is a vacuum. Figure 2.11 illustrates an example of heat transfer by radiation.

Heat Transfer by Conduction

Figure 2.12 shows the direction of heat flow in a solid concrete masonry wall by conduction. Warm, excited molecules transfer some of their vibrational energy to adjacent, cooler, less excited molecules. The flow of heat energy is always away from the warmer region to the cooler region. The rate of heat flow by conduction is directly proportional to the thermal conductivity, k, of the material and the thermal gradient, Δt.

Thermal Properties of Masonry Materials

The intrinsic properties of a material determine its thermal characteristics. Two properties which are influential in determining a

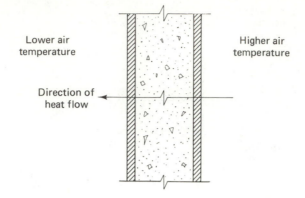

Figure 2.12 Heat flow through a concrete masonry wall by conduction

material's thermal performance are conductivity (the rate of heat transfer through the material) and heat capacity (the ability of a material to store heat, dependent on the specific heat and density of the material). Dense, massive materials, such as masonry elements, are very good at storing heat. Passive solar building designs are based on the principle that massive elements ("thermal mass") in a room will absorb heat during the day when exposed to solar radiation. At night, when the temperature inside the room starts to drop, the heat stored in the mass is slowly released. Masonry construction is an efficient method of incorporating mass into a building, and thus improving the overall energy performance of the building. Masonry can be incorporated into a building design in the form of interior walls, exterior walls, fireplaces, planters, floor surfaces, and benches, as well as many other applications [2.8].

Insulation

Insulators, such as wood, dry soil, and plastic foam, have a low coefficient of thermal conductivity and resist the flow of heat across the material. In general, the less dense a material is, the better insulator it is.

The resistance of concrete masonry walls can be improved through the use of insulation. There are several types currently available and the selection of a particular type depends on several factors. If the concrete masonry is to be partially grouted or requires an interior finish, there are insulation materials which can be applied to the inside face of the concrete masonry.

Insulation intended for application inside the cells of the concrete masonry unit consists of three basic types:

1. *Granular:*
   a. *Perlite* (glassy volcanic rock expanded by heating). May be treated with silicone to increase resistance to water penetration.
   b. *Vermiculite* (hydrated magnesium aluminum iron silicate with a foliated structure) is expanded by heating and may be treated for water repelling.
2. *Foamed-in-place:* plastic materials together with foaming and hardening agents combined while being forced under air pressure into the cells of the concrete masonry. It may also be sprayed over exposed wall surfaces. The foam will expand and cure in place.
3. *Foamed:* plastic substances expanded into foam by mixing with air, carbon dioxide, fluorocarbons, or other gaseous media and molded into rigid shapes. Gaseous media are trapped in the closed cells of the insulation, which comprise up to 90 percent of the total volume.

Granular and formed-in-place insulation is placed in the cells after the wall is built.

The foamed type of insulation placed into the cells of the concrete masonry units prior to being laid up into a wall allows this type of insulation to be used in grouted masonry construction. Figure 2.13 shows these preinsulated concrete masonry units.

The most common type of insulation used in fully grouted concrete masonry construction is a rigid polystyrene foam insulation board attached to the masonry by fasteners or adhesives. Furring strips may be factory-installed U-shaped galvanized steel or site-installed wood furring strips shot to the wall. The finished wall surface may then be applied over the insulation and nailed or screwed into the furring strips. Figure 2.14 shows the installation of this type of insulation.

The application of these insulation techniques can result in concrete masonry construction which possesses the thermal resistance values mandated by the various building codes.

Figure 2.13 Preinsulated masonry units for grouted or ungrouted construction (Courtesy of Korfil, Inc.)

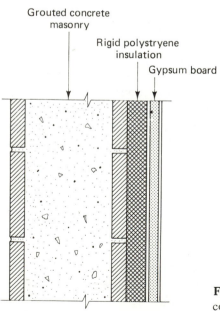

Grouted concrete masonry

Rigid polystryene insulation

Gypsum board

Figure 2.14 Insulation for solidly grouted concrete masonry

Steady-State Heat Flow Calculations

The measure of heat transfer by conduction is the most common method of evaluating the thermal performance of a material. The thermal conductivity is the time rate of heat flow (Btu per hour) through a material of unit area (1 square foot) and unit thickness (1 inch) for a unit temperature differential (1 degree F). Thus, the units of thermal conductivity are expressed in terms of Btu/hr-ft^2-°F/in.

Because walls are usually constructed of materials with standard thicknesses, such as grouted 8-in. concrete masonry, a term normalized by dividing the thermal conductivity by the material thickness, t, is often more convenient to use in calculating heat flow. The thermal conductance, C, is

$$C = \frac{k}{t} \tag{2.4}$$

The reciprocal of the thermal conductance is the thermal resistance, R, and is equal to

$$R = \frac{1}{C}$$

$$= \frac{t}{k} \tag{2.5}$$

The R-value, expressed in units of °F-ft^2-hr/Btu is a measure of the resistance to heat flow of a material with a given thickness.

Similar to the thermal conductivity value, k, a measure of resistance to heat flow through a material per inch of thickness is given by the thermal resistivity, r. The thermal resistivity is related to the thermal resistance by

$$r = \frac{R}{t} \tag{2.6}$$

The total resistance to heat transfer of a building assembly, such as a wall, floor, or ceiling assembly, is equal to the sum of the resistances of the individual layers:

$$R_T = R_1 + R_2 + R_3 + \cdots + R_n \quad (R = \text{°F-ft}^2\text{-hr/Btu}) \tag{2.7}$$

The total resistance of the assembly includes the resistance of the air films on the inside and outside of the construction.

The overall coefficient of heat transfer of a construction assembly is the U-value, and is equal to the inverse of the sum of the resistances of the assembly:

$$U = \frac{1}{R_T} \qquad (U = \text{Btu/hr-ft}^2\text{-}°\text{F}) \qquad (2.8)$$

This methodology for calculating total heat transfer is appropriate for walls where the layers of the assembly have uniform conductances throughout.

In many instances, a construction assembly is composed of elements in such a manner that there are parallel heat flow paths through the assembly with different conductances. If there is no significant lateral heat flow between the paths, each assembly with a different heat flow path can be assumed to extend from inside to out. The overall transmittance of each path may be calculated using the equation (2.8), and an average U-value calculated based on the percentage of the construction that consists of each type of heat flow path:

$$U_{av} = a(U_a) + b(U_b) + \cdots + n(U_n) \qquad (2.9)$$

where a, b, $\cdots$, n are the percentages of each heat flow path, in a typical unit area, where the overall transmittances of each path are $U_a, U_b, \cdots, U_n$ [2.9].

A concrete masonry wall is a complex three-dimensional assembly of block material, mortar, grout, void, and perhaps insulation. Precise calculations of heat transfer can be very difficult, and a detailed analysis requires heat transfer algorithms not commonly used, such as the series parallel heat flow equations. These calculations take into account lateral heat flow between parallel heat flow paths. The most precise thermal evaluation of a material is actual physical testing.

The *Concrete Masonry Design Manual* [2.10], published by the Concrete Masonry Association of California and Nevada, contains a simplified method for calculating the U-value for a masonry wall. The method consists of a set of tables which list the percentage of grouted and ungrouted block for typical constructions, and a set of tables listing the R- and U-values for common block wall constructions. It is then simply a matter of finding an average U-value for the construction. This method is based on the theory represented in Equation (2.9), and assumes no lateral heat flow paths.

EXAMPLE

This example uses the methodology from the *Concrete Masonry Design Manual*, based on parallel heat flow.

A wall is constructed of 8 in. × 8 in. × 16 in., 100-pcf, two-cell concrete block with 140-pcf grout. Reinforcement is placed 48 in. on center horizontally and 32 in. on center vertically. Five typical constructions are calculated; in all cases, the cells with reinforcing steel are grouted:

1. All cells, with and without steel, are grouted.
2. The unreinforced cells are left empty.
3. The unreinforced cells are filled with perlite insulation.
4. The unreinforced cells are filled with polystyrofoam insulation.
5. All cells are grouted, wall has 1 in. of polystyrene foam and sheetrock on the inside.

SOLUTION

From Table 2.8:

$$\text{Percentage of wall that is grouted} = 37.5\%$$

$$\text{Percentage of wall that is ungrouted} = 62.5\%$$

From Table 2.9:

(1) The *U*-value for grouted block   = 0.45

(2) The *U*-value for block with empty cells   = 0.39

(3) The *U*-value for block with perlite   = 0.15

(4) The *U*-value for block with polyurethane = 0.13

(5) The *U*-value for grouted block with 1 in. of polystyrofoam and sheetrock   = 0.13

Using Equation (2.9) gives us

$$U_{av} = (\%\text{ grouted}) \times (U\text{-value grouted}) + (\%\text{ ungrouted})$$

$$\times (U\text{-value ungrouted})$$

TABLE 2.8   Block 8 in. High, 16 in. Long, Two Cells [2.10]

% Ungrouted

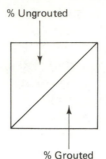

% Grouted

		Vertical Steel Spacing (in.)					
		48	40	32	24	16	8
Joint Reinf. Only[a]		83.3 / 16.7	80.0 / 20.0	75.0 / 25.0	66.7 / 33.3	50.0 / 50.0	0 / 100
Horizontal Steel Spacing[c] (in.)	104[b]	76.9 / 23.1	73.8 / 26.2	69.2 / 30.8	61.5 / 38.5	46.2 / 53.8	0 / 100
	48	69.4 / 30.6	66.7 / 33.3	62.5 / 37.5	55.6 / 44.4	41.7 / 58.3	0 / 100
	40	66.7 / 33.3	64.0 / 36.0	60.0 / 40.0	53.3 / 46.7	40.0 / 60.0	0 / 100
	32	62.5 / 37.5	60.0 / 40.0	56.2 / 43.8	50.0 / 50.0	37.5 / 62.5	0 / 100
	24	55.6 / 44.4	53.3 / 46.7	50.0 / 50.0	44.4 / 55.6	33.3 / 66.7	0 / 100
	16	41.7 / 58.3	40.0 / 60.0	37.5 / 62.5	33.3 / 66.7	25.0 / 75.0	0 / 100
	8	0 / 100	0 / 100	0 / 100	0 / 100	0 / 100	0 / 100

[a]Reinforcing steel grouted in vertical cells only. Horizontal steel in joints. Ungrouted cells are uninterrupted.

[b]Wall 8 ft 8 in. high, top 8 in. bond beam reinforced and grouted. Joints reinforcing below bond beam, vertical cells are uninterrupted.

[c]Horizontal bond beam 8 in. high.

TABLE 2.9 Heating R- and U-Values for Concrete Masonry Walls[a] [2.10]

A. 8-in. Concrete Block; Solid Grouted

Surface Treatment	Solid Grout: Density 90 pcf		Solid Grout: Density 100 pcf		Solid Grout: Density 110 pcf		Solid Grout: Density 120 pcf		Solid Grout: Density 140 pcf		Density of Block (pcf)
	R	U	R	U	R	U	R	U	R	U	
Block exposed both sides	3.23	0.31	3.07	0.33	2.87	0.35	2.70	0.37	2.34	0.43	90
	3.09	0.32	2.95	0.34	2.75	0.36	2.59	0.39	2.23	0.45	100
	2.92	0.34	2.78	0.36	2.59	0.39	2.43	0.41	2.09	0.48	110
	2.76	0.36	2.63	0.38	2.45	0.41	2.30	0.43	1.97	0.51	120
	2.58	0.39	2.46	0.41	2.30	0.44	2.16	0.46	1.84	0.54	130
Plain gypsum board on $\frac{3}{4}$-in. furring	4.65	0.22	4.49	0.22	4.29	0.23	4.12	0.24	3.76	0.27	90
	4.51	0.22	4.37	0.23	4.17	0.24	4.01	0.25	3.65	0.27	100
	4.34	0.23	4.20	0.24	4.01	0.25	3.85	0.26	3.51	0.29	110
	4.18	0.24	4.05	0.25	3.87	0.26	3.72	0.27	3.39	0.30	120
	4.00	0.25	3.88	0.26	3.72	0.27	3.58	0.28	3.26	0.31	130
Foil backed gypsum board on $\frac{3}{4}$-in. furring	6.22	0.16	6.06	0.16	5.86	0.17	5.69	0.18	5.33	0.19	90
	6.08	0.16	5.94	0.17	5.74	0.17	5.58	0.18	5.22	0.19	100
	5.91	0.17	5.77	0.17	5.58	0.18	5.42	0.18	5.08	0.20	110
	5.75	0.17	5.62	0.18	5.44	0.18	5.29	0.19	4.96	0.20	120
	5.57	0.18	5.45	0.18	5.29	0.19	5.15	0.19	4.83	0.21	130
1-in. fiberglass board and $\frac{1}{2}$-in. gypsum board	7.68	0.13	7.52	0.13	7.32	0.14	7.15	0.14	6.79	0.15	90
	7.54	0.13	7.40	0.14	7.20	0.14	7.04	0.14	6.68	0.15	100
	7.37	0.14	7.23	0.14	7.04	0.14	6.88	0.15	6.54	0.15	110
	7.21	0.14	7.08	0.14	6.90	0.14	6.75	0.15	6.42	0.16	120
	7.03	0.14	6.91	0.14	6.75	0.15	6.61	0.15	6.29	0.16	130
1-in. polystyrofoam board and $\frac{1}{2}$-in. gypsum board	8.68	0.12	8.52	0.12	8.32	0.12	8.15	0.12	7.79	0.13	90
	8.54	0.12	8.40	0.12	8.20	0.12	8.04	0.12	7.68	0.13	100
	8.37	0.12	8.23	0.12	8.04	0.12	7.88	0.13	7.54	0.13	110

TABLE 2.9 (cont.)

A. 8-in. Concrete Block; Solid Grouted

Surface Treatment	Solid Grout: Density 90 pcf		Solid Grout: Density 100 pcf		Solid Grout: Density 110 pcf		Solid Grout: Density 120 pcf		Solid Grout: Density 140 pcf		Density of Block (pcf)
	R	U	R	U	R	U	R	U	R	U	
	8.21	0.12	8.08	0.12	7.90	0.13	7.55	0.13	7.42	0.13	120
	8.03	0.12	7.91	0.13	7.75	0.13	7.61	0.13	7.29	0.14	130
1-in. polyurethane foam and ½-in. gypsum board	9.93	0.10	9.77	0.10	9.57	0.10	9.40	0.11	9.04	0.11	90
	9.79	0.10	9.65	0.10	9.45	0.11	9.29	0.11	8.93	0.11	100
	9.62	0.10	9.48	0.11	9.29	0.11	9.13	0.11	8.79	0.11	110
	9.46	0.11	9.33	0.11	9.15	0.11	9.00	0.11	8.67	0.12	120
	9.28	0.11	9.16	0.11	9.00	0.11	8.86	0.11	8.54	0.12	130

B. 8-in. Concrete Block; Cells Insulated

Surface Treatment	No Insulation Fill (Void)		Vermiculite Fill		Perlite Fill		Urea-Formaldehyde Foam		Polyurethane Foam		Density of Block (pcf)
	R	U	R	U	R	U	R	U	R	U	
Block exposed both sides	2.70	0.37	6.63	0.15	7.10	0.14	8.16	0.12	8.47	0.12	90
	2.58	0.39	6.24	0.16	6.65	0.15	7.56	0.13	7.82	0.13	100
	2.43	0.41	5.67	0.18	6.01	0.17	6.72	0.15	6.92	0.14	110
	2.30	0.44	5.15	0.19	5.41	0.18	5.97	0.17	6.12	0.16	120
	2.16	0.46	4.53	0.22	4.73	0.21	5.13	0.19	5.24	0.19	130

Material	Temp										
Plain gypsum board on $\frac{3}{4}$-in. furring	90	4.12	0.24	8.05	0.12	8.52	0.12	9.58	0.10	9.89	0.10
	100	4.00	0.25	7.66	0.13	8.07	0.12	8.98	0.11	9.24	0.11
	110	3.85	0.26	7.09	0.14	7.43	0.13	8.14	0.12	8.34	0.12
	120	3.72	0.27	6.57	0.15	6.83	0.15	7.39	0.14	7.54	0.13
	130	3.58	0.28	5.95	0.17	6.15	0.16	6.55	0.15	6.66	0.15
Foil backed gypsum board on $\frac{3}{4}$-in. furring	90	5.69	0.18	9.62	0.10	10.09	0.10	11.15	0.09	11.46	0.09
	100	5.57	0.18	9.23	0.11	9.64	0.10	10.55	0.09	10.81	0.09
	110	5.42	0.18	8.66	0.12	9.00	0.11	9.71	0.10	9.91	0.10
	120	5.29	0.19	8.14	0.12	8.40	0.12	8.96	0.11	9.11	0.11
	130	5.15	0.19	7.52	0.13	7.72	0.13	8.12	0.12	8.23	0.12
1-in. fiberglass board and $\frac{1}{2}$-in. gypsum board	90	7.15	0.14	11.08	0.09	11.55	0.09	12.61	0.08	12.92	0.08
	100	7.03	0.14	10.69	0.09	11.10	0.09	12.01	0.08	12.37	0.08
	110	6.88	0.15	10.12	0.10	10.46	0.10	11.17	0.09	11.37	0.09
	120	6.75	0.15	9.60	0.10	9.86	0.10	10.42	0.10	10.57	0.09
	130	6.61	0.15	8.98	0.11	9.18	0.11	9.58	0.10	9.69	0.10
1-in. polystyrofoam board and $\frac{1}{2}$-in. gypsum board	90	8.15	0.12	12.08	0.08	12.55	0.08	13.61	0.07	13.92	0.07
	100	8.03	0.12	11.69	0.09	12.10	0.08	13.01	0.08	13.27	0.08
	110	7.88	0.13	11.12	0.09	11.46	0.09	12.17	0.08	12.37	0.08
	120	7.75	0.13	10.60	0.09	10.86	0.09	11.42	0.09	11.57	0.09
	130	7.61	0.13	9.98	0.10	10.18	0.10	10.58	0.09	10.69	0.09
1-in. polyurethane foam and $\frac{1}{2}$-in. gypsum board	90	9.40	0.11	13.33	0.07	13.80	0.07	14.86	0.07	15.17	0.07
	100	9.28	0.11	12.94	0.08	13.35	0.07	14.26	0.07	14.52	0.07
	110	9.13	0.11	12.37	0.08	12.71	0.08	13.42	0.07	13.62	0.07
	120	9.00	0.11	11.85	0.08	12.11	0.08	12.67	0.08	12.82	0.08
	130	9.86	0.11	11.23	0.09	11.43	0.09	11.83	0.08	11.94	0.08

[a]Includes surface air films; air to air.

TABLE 2.10   Average U-Values for Example

Wall Construction	U-Value Reinforced Block[a]	Percent Reinforced Block	U-Value Unreinforced Block[a]	Percent Unreinforced Block	Average U-Value[a]
(1) All cores grouted	(0.45	× 0.375) +	(0.45	× 0.625) =	0.45
(2) Ungrouted cores are empty	(0.45	× 0.375) +	(0.39	× 0.625) =	0.41
(3) Ungrouted cores with perlite	(0.45	× 0.375) +	(0.15	× 0.625) =	0.26
(4) Ungrouted cores with polyurethane	(0.45	× 0.375) +	(0.13	× 0.625) =	0.25
(5) All cores grouted, 1-in. interior polystyrofoam board, $\frac{1}{2}$-in. gypsum board	(0.13	× 0.375) +	(0.13	× 0.625) =	0.13

[a]These values include air films.

The resulting average U-values for each case are shown in Table 2.10.

Non-Steady-State Heat Flow

An important characteristic of masonry materials is their ability to store heat. The heat capacity of a material is based on its properties of specific heat, density, thermal conductivity, and thickness.

Steady-state heat flow calculations are not able to take heat capacity of a material into account in predicting thermal performance. However, analysis of masonry walls with an hourly simulation computer program will model the wall's thermal behavior based on non-steady-state heat flow algorithms. This is much more sophisticated than steady-state methods of analyzing the thermal performance of a material, because it takes into account hourly changes in temperature, and is able to incorporate the lag time of heat transfer through a massive material. Such analysis can be used to show that, due to its heat capacity properties, a masonry wall performs thermally in a much different way than a light frame wall.

For example, in the "Energy Conservation Standards for New Residential Buildings," for the State of California [2.11], computer analysis on a typical tract residence compared the thermal performance of a house with light frame walls and one with masonry walls. The result from this analysis was that the R-value of a masonry exterior wall could be lower than the R-value of a light frame wall and the house would still use the same amount of energy.

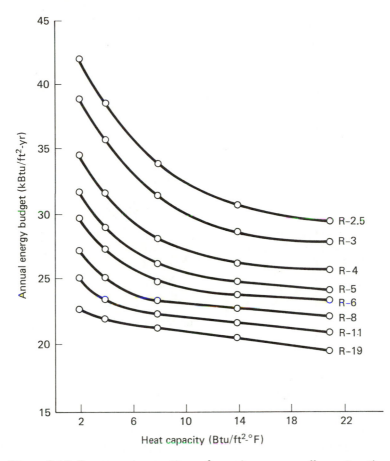

Figure 2.15 Energy consumption of varying mass wall construction, Sacramento, California [2.11]

Based on this computer analysis, the graph in Figure 2.15 was generated for masonry exterior walls having varying R-values and heat capacities for Sacramento, California. The graph shows that for each R-value wall, the annual energy consumption of the house decreases as the heat capacity of the wall increases.

The thermal performance of masonry is very climate dependent. In general, if the inside temperature of the building floats off the thermostat setting of the house, addition of thermal mass, in the form of masonry materials for example, will reduce the amount of energy used by the building. If the house was in a climate dominated by cooling loads, the massive house performed much better over the

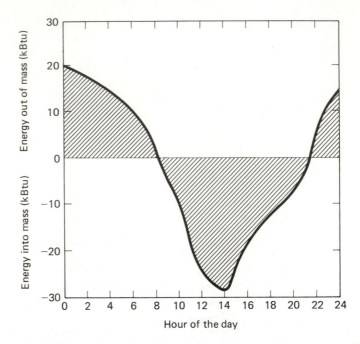

Figure 2.16 Exterior mass wall heat flow diagram, Sacramento, California; typical summer day [2.11]

course of the year. If the climate was dominated by heating loads, the massive house did not reduce the conditioning load as dramatically.

Figure 2.16 is an hourly heat flow diagram of an exterior mass wall for Sacramento, California, on a typical summer day. This diagram, generated by computer simulation of the non-steady-state heat flow through the wall, shows that heat is leaving the mass wall at night, when air temperatures are low, and that heat is being absorbed by the mass wall during the day, when air temperatures are high. In this case, the heat storage capacity of masonry is used to advantage in keeping the building cool.

3

STRUCTURAL MECHANICS

3.1 GENERAL

Reinforced concrete masonry has traditionally been designed according to an elastic theory which assumes that stress is proportional to strain. It has also been assumed that strain varies linearly from the neutral axis. Working stress design (WSD) can predict the behavior of masonry provided that the stress level is less than the proportional limit of the material. WSD does not accurately predict the ultimate load-carrying capacity of a masonry member because it does not accurately model stresses when strains are in the inelastic range.

The engineer must be able to estimate the ultimate load-carrying capacity of a masonry member. This is necessary to produce a safe and economical design. State-of-the-art design methods in all areas of earthquake engineering seek to establish accurate potential modes of failure of a structural member and predict the capacity of the member at first yield as well as at failure. The strength design method (SDM) is accepted in reinforced concrete design because it enables the designer to estimate member strength associated with a specific failure mode.

This chapter presents the mechanics of concrete masonry. It

develops the basic tools necessary to enable the engineer to describe the behavior of concrete masonry at all strain levels. At small levels of masonry strain, the stress-strain relationship in the concrete masonry is linear and an elastic approach is appropriate. These strain levels are usually associated with serviceability design considerations such as cracking. At larger strain levels the stress-strain relationship is nonlinear.

3.2 MEMBERS SUBJECTED TO AXIAL COMPRESSION AND TENSION

A fundamental assumption in reinforced concrete masonry theory is that the heterogeneous combination of concrete masonry, mortar, grout, and steel is able to function as a single unit because of strain compatibility. This concept assures that in a uniformly compressed section, the strain in the masonry (ϵ_m) is equal to the strain in the steel (ϵ_s) and permits the following theoretical development used to describe axial compression and tension in a reinforced masonry member.

Figure 3.1 superimposes an idealized stress-strain curve for concrete masonry on that of steel. The stress-strain curves show a straight-line relationship up to the proportional limit of each of the materials. The proportional limit of the steel is assumed to coincide

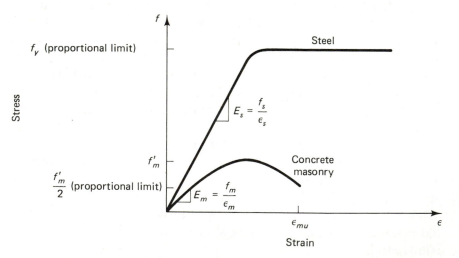

Figure 3.1 Idealized stress-strain curves for steel and concrete masonry

with the yield point, whereas in concrete masonry, the proportional limit is taken as being approximately one-half of the ultimate compressive strength of the masonry, f_m'. Tests on unconfined masonry specimens indicate that the compressive strength of the masonry is reached at a strain of approximately 0.0015 [3.1].

From Figure 3.1 it can be seen that the concrete masonry is able to take additional load as the strain increases up to the ultimate compressive strength, f_m'. After that point, increasing the strain results in a decrease in the capacity of the concrete masonry until the section fails at the ultimate strain, ϵ_{mu}.

For the range of strains where the modulus of elasticity of concrete masonry, E_m, and the modulus of elasticity of steel, E_s, are constant, the modular ratio, n, is also constant and is given by

$$n = \frac{E_s}{E_m} \tag{3.1}$$

The stresses in the masonry, f_m, and the steel, f_s, are then given by

$$f_m = E_m \epsilon_m \tag{3.2a}$$

$$f_s = E_s \epsilon_s \tag{3.2b}$$

and the two stresses can be related to each other through the modular ratio

$$f_s = n f_m \tag{3.3}$$

If it is assumed that secondary lateral deflections caused by an eccentrically applied load are ignored, the axial load, P, on a member stressed in the elastic range can now be expressed as the sum of the axial load resisted by the concrete masonry and the axial load resisted by the steel.

$$P = f_m A_m + f_s A_s$$
$$= f_m (A_m + n A_s) \tag{3.4}$$

where A_m = net area of concrete masonry (gross area of the section minus the area of steel)
A_s = area of steel reinforcing
P = axial load in compression

The term $(A_m + n A_s)$ is called the *transformed area* and represents a fictitious masonry area stressed to f_m by the axial load P. The transformed area is constructed by assuming that a given area of steel can resist n times the force that an equal area of concrete masonry is capable of resisting. The actual steel area is then transformed into a fictitious area of masonry, as shown in Figure 3.2. Figure 3.2 also shows that Equation (3.4) can be rewritten in a more convenient form:

$$P = f_m [A_g + (n - 1) A_s] \qquad (3.5)$$

where A_g is the gross area of the section. Equation (3.5) will give the designer the allowable axial compressive load if f_m is a working (allowable) stress within the elastic range (i.e., $f_m \leqslant \frac{1}{2} f_m'$) and if the modular ratio is known.

As additional load is applied to the member, the strain in the concrete masonry and the strain in the steel both increase at the same rate because of the assumed perfect bond between the two

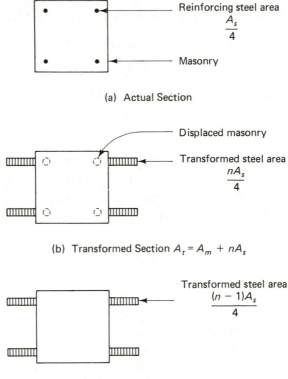

(a) Actual Section

(b) Transformed Section $A_t = A_m + nA_s$

(c) Transformed Section $A_t = A_g + (n - 1)A_s$

Figure 3.2 Transformed section in axial compression

materials. The loading on the section can be increased, in fact, to the point where the member fails in compression; however, the type of the compression failure depends on the yield stress of the steel and the ultimate compressive strain of the concrete masonry.

Knowledge of the ultimate compressive strain of concrete masonry (ϵ_{mu}) is important because the member is assumed to have failed when this strain level has been reached. Once the masonry has failed by crushing, the reinforcing steel is no longer confined and can take no compressive load. Extensive tests on unconfined concrete members indicate that the material can undergo ultimate strains up to 0.007 [3.2]. An accepted conservative value for the maximum unconfined compressive strain in concrete is 0.003. Such a comprehensive research effort has not yet been conducted in concrete masonry; however, a review of available test data indicates that concrete masonry can support load with strains as high as 0.005 [3.1]. An appropriate value for the ultimate compressive strain in masonry, ϵ_{mu}, of 0.002 is reasonable for design purposes, although values of ϵ_{mu} as high as 0.0025 to 0.003 have been proposed [3.3].

The yield stress of the steel, f_y, is also important, for it allows the designer to establish the yield strain of the steel, ϵ_y:

$$\epsilon_y = \frac{f_y}{E_s} \tag{3.6}$$

If the yield strain of the steel is less than the ultimate compressive strain of the masonry (i.e., $\epsilon_y < \epsilon_{mu}$), the steel will yield before the concrete masonry fails in compression. This implies that the maximum load which the steel can resist is reached when the stress in the steel is equal to its yield stress. If we assume no strain hardening in the reinforcement, continued loading results in the plastic straining of the steel, with the additional load being resisted only by the masonry. The load on the section can now be increased until the masonry fails in compression. This is the case for Grade 40 reinforcing, where f_y is equal to 40 ksi and ϵ_y is equal to 0.0014.

When the yield strain of the steel is equal to, or nearly equal to the ultimate compressive strain of the masonry (i.e., $\epsilon_y = \epsilon_{mu}$), the section will fail as the strain approaches ϵ_{mu}. This is the case for Grade 60 reinforcing ($f_y = 60$ ksi, $\epsilon_y = 0.0021$). If the ultimate compressive strain in the masonry is reached before the steel yields (i.e., $\epsilon_y > \epsilon_{mu}$), a crushing failure comparable to that described for the case where $\epsilon_y = \epsilon_{mu}$ will occur.

If it is assumed that the selection of steels is limited to those where $\epsilon_y \leqslant \epsilon_{mu}$, the nominal compressive strength, P_n, is given by

$$P_n = 0.85 f_m' A_m + f_y A_s \qquad (3.7)$$

It can be seen by comparing Equation (3.7) with the equations developed for the elastic range [e.g., Equation (3.4) or (3.5)] that the modular ratio, n, is no longer required to determine the capacity of the section.

Although the ultimate compressive strength of the concrete masonry is assumed to be f_m', this value is obtained from prisms loaded at a fairly high rate of strain. Actual loading rates on real members are typically much slower, and the ultimate compressive stress is, therefore, somewhat less than the tested f_m'. Tests on concrete members indicate that the maximum reliable concrete strength is approximately $0.85 f_c'$ [3.4]. Similar behavior in concrete masonry should be expected, and the modified ultimate compressive strength is assumed to be equal to $0.85 f_m'$.

The tensile strength of concrete masonry, like that of concrete, is small compared to its compressive strength. For this reason, reinforced masonry members are seldom designed in a manner that depends on substantial tensile strength from the masonry; however, at low levels of strain, the accurate prediction of member behavior can only be made by considering the tensile resistance provided by the masonry.

If the tensile forces are of a magnitude such that the tensile strength of the concrete masonry is not exceeded, Equation (3.5) can be used to determine the axial load when both the steel and masonry will behave elastically:

$$P_t = f_{mt}[A_g + (n - 1)A_s] \qquad (3.8)$$

where f_{mt} is the tensile stress in concrete masonry.

As the load is increased, the concrete masonry cracks and the steel must resist the entire tensile load. The load on the member is now expressed in terms of the steel alone:

$$P = f_s A_s \qquad (3.9)$$

where f_s is the steel stress.

As the load is increased to the point where $f_s = f_y$, the small elastic strains previously present in the steel give way to permanent inelastic deformations. The nominal strength of the member in tension has now been reached and is given by

$$P_n = f_y A_s \qquad (3.10)$$

EXAMPLE 3.1

A compression test has been run on a grouted 16 × 16 in. concrete masonry pilaster to generate the stress-strain curve shown in Figure E3.1-1a. A second test was conducted on a sample of Grade 40 reinforcing steel, and the stress-strain curve shown in Figure E3.1-1b was obtained. A 16 × 16 in. square reinforced pilaster was then constructed from material similar to that on which the tests were conducted, and is shown in Figure E3.1-2.

Using the information supplied by the stress-strain curves, determine:

1. The stress and strain in the concrete masonry and the steel at the proportional limit of the masonry
2. The axial load, P, to stress the masonry to the proportional limit of the masonry
3. The axial load, P, at the maximum masonry stress
4. The axial load, P_y, to yield the steel
5. The axial load, P_n, at the ultimate strain

SOLUTION TO PART 1

An examination of the stress-strain curve for the concrete masonry shows that the stress is linearly proportional up to approximately $\epsilon_m = 0.0005$. This defines the strain at the proportional limit. From the two stress-strain curves in Figure E3.1-1, the stress of the steel, f_s, and masonry, f_m, for a strain of 0.0005 are

$$f_s = 14.50 \text{ ksi}$$
$$f_m = 0.703 \text{ ksi}$$

SOLUTION TO PART 2

Area of steel = A_s = 4(0.31 in.^{2}) = 1.24 in.^{2}

Area of masonry = A_m = (15.63 in.)^2 − 1.24 in.^{2} = 243.1 in.^{2}

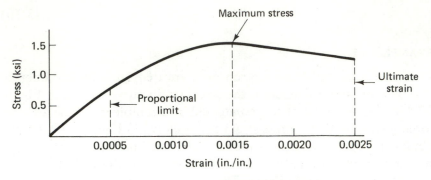

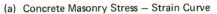

(a) Concrete Masonry Stress – Strain Curve

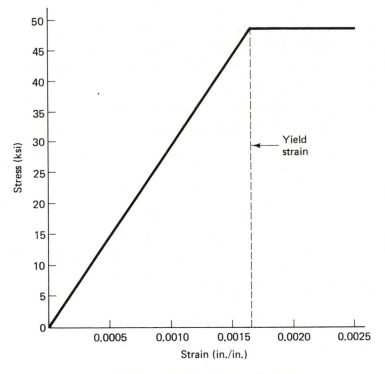

(b) Reinforcing Steel Stress – Strain Curve

Figure E3.1-1 Material stress-strain curves

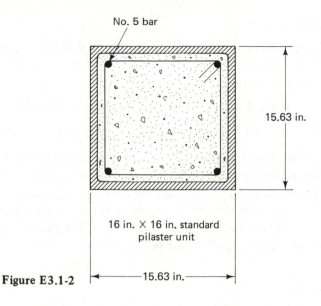

No. 5 bar

15.63 in.

16 in. × 16 in. standard
pilaster unit

Figure E3.1-2   ⟵————15.63 in.————⟶

$P = P_{\text{masonry}} + P_{\text{steel}}$

$= 0.703 \text{ ksi}(243.1 \text{ in.}^2) + 14.5 \text{ ksi}(1.24 \text{ in.}^2)$

$= 170.9 \text{ kips} + 18.0 \text{ kips} = 188.9 \text{ kips}$

SOLUTION TO PART 3

The maximum masonry stress is $f'_m = 1.50$ ksi at a strain of $\epsilon_m = 0.0015$. The corresponding steel stress at a strain of 0.0015 is $f_s = 43.5$ ksi.

$P = P_{\text{masonry}} + P_{\text{steel}}$

$= 1.50 \text{ ksi}(243.1 \text{ in.}^2) + 43.5 \text{ ksi}(1.24 \text{ in.}^2)$

$= 364.7 \text{ kips} + 55.0 \text{ kips} = 419.7 \text{ kips}$

SOLUTION TO PART 4

The steel yields at a strain $\epsilon_y = 0.00165$ at a stress $f_y = 47.9$ ksi. The masonry stress at this point is $f_m = 1.45$ ksi.

$P = 1.45 \text{ ksi}(243.1 \text{ in.}^2) + 47.9 \text{ ksi}(1.24 \text{ in.}^2)$

$= 352.5 \text{ kips} + 59.4 \text{ kips} = 411.9 \text{ kips}$

SOLUTION TO PART 5

The ultimate strain in the masonry is $\epsilon_{mu} = 0.0025$ and the stress is $f_m = 1.2$ ksi. The steel stress remains $f_s = 47.9$ ksi because the strain is above the yield strain.

$$P = 1.2 \text{ ksi}(243.1 \text{ in.}^2) + 47.9 \text{ ksi}(1.24 \text{ in.}^2)$$

$$= 291.7 \text{ kips} + 59.4 \text{ kips} = 351.1 \text{ kips}$$

Note that the contribution of the steel to the axial load in Parts 2 to 5 varies as follows: 9.5, 13.1, 14.4, and 16.9 percent. It is left to the reader to study the sensitivity of P to A_s, f'_m, and f_y. Also note that this column exhibits the desirable behavior of steel yielding prior to masonry failure because $\epsilon_y = 0.00165$ is less than $\epsilon_{mu} = 0.0025$. Do the areas A_m or A_s alter this behavior?

EXAMPLE 3.2

Consider the same reinforced concrete masonry pilaster shown in Figure E3.1-2. Assume that $f'_m = 1.5$ ksi and $f_y = 40$ ksi. The modulus of elasticity of the concrete masonry is $E_m = 1000f'_m$ and that of the steel is $E_s = 29,000$ ksi.

Using the elastic theory presented in this chapter, determine:

1. The stress and strain in the masonry and the steel at the proportional limit of the concrete masonry
2. The axial load, P, to stress the masonry to the proportional limit of the masonry
3. The axial load, P, at the maximum masonry stress, f'_m
4. The axial load, P, to stress the masonry to the proportional limit of $E_m = 600f'_m = 900$ ksi
5. The axial load at the maximum masonry stress, assuming that $E_m = 900$ ksi

SOLUTION TO PART 1

The masonry stress at the proportional limit is approximately $f_m = 0.5f'_m = 0.75$ ksi.

$$E_m = 1000f'_m$$

$$= 1000(1.5 \text{ ksi}) = 1500 \text{ ksi}$$

$$\epsilon_m = \frac{f_m}{E_m}$$

$$= \frac{0.75}{1500} = 0.0005$$

$$f_s = E_s \epsilon_s$$

$$= 29,000 \text{ ksi}(0.0005) = 14.5 \text{ ksi}$$

SOLUTION TO PART 2

$$n = \frac{E_s}{E_m}$$

$$= \frac{29,000}{1500} = 19.33$$

$$A_g = (15.63 \text{ in.})^2 = 244.3 \text{ in.}^2$$

$$P = f_m [A_g + (n - 1)A_s]$$

$$= 0.75 \text{ ksi}[244.3 \text{ in.}^2 + (19.33 - 1)1.24 \text{ in.}^2] = 200.3 \text{ kips}$$

SOLUTION TO PART 3

$$P = 1.5 \text{ ksi}[244.3 \text{ in.}^2 + (19.33 - 1)1.24 \text{ in.}^2] = 400.5 \text{ kips}$$

$$\epsilon_m = \frac{f_m'}{E_m}$$

$$= \frac{1.5}{1500} = 0.001$$

SOLUTION TO PART 4

$$n = \frac{29,000}{900} = 32.2$$

$$P = 0.75 \text{ ksi}[244.3 \text{ in.}^2 + (32.2 - 1)1.24 \text{ in.}^2] = 212.3 \text{ kips}$$

SOLUTION TO PART 5

$$P = 1.5 \text{ ksi}[244.3 \text{ in.}^2 + (32.2 - 1)1.24 \text{ in.}^2] = 424.5 \text{ kips}$$

EXAMPLE 3.3

Consider the same reinforced concrete masonry pilaster shown in Figure E3.1-2 and assume that it possesses the same material properties as those given in Example 3.2. Using the ultimate strength theory presented in this chapter, determine the nominal ultimate load, P_n.

SOLUTION

$$P = 0.85 f'_m A_m + f_y A_s$$

$$= 0.85(1.5 \text{ ksi})243.1 \text{ in.}^2 + 40 \text{ ksi}(1.24 \text{ in.}^2) = 359.6 \text{ kips}$$

Figure E3.3 plots the load/deformation curves for the pilasters analyzed in Examples 3.1 through 3.3. This curve based on the stress-strain curves from Example 3.1 has a straight-line portion up to the proportional limit of the masonry, and then the nonlinear behavior of the masonry results in the nonlinear curve for strains greater than

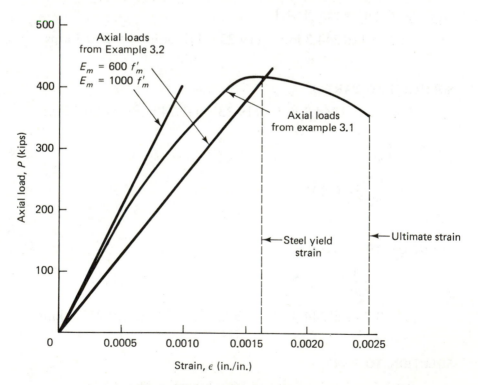

Figure E3.3 Load/deformation curve

$E_m = 0.0005$. The linear behavior of the curves from Example 3.2 results from the application of the elastic theory through the total loading range.

3.3 MEMBERS SUBJECTED TO FLEXURE

The structural mechanics developed for a reinforced concrete masonry section in flexure attempt to model its true behavior. This allows the designer to understand and predict the behavior of the masonry member. The development of this theory requires the acceptance of several basic assumptions, the implications of which the designer must understand. The four basic assumptions made in developing a flexural theory for reinforced concrete masonry are:

1. Plane sections before bending remain plane after being loaded.
2. The stress-strain relationship for the masonry is known.
3. The tensile strength of the masonry is neglected except at very low stress levels.
4. The stress-strain relationship for the steel is known.

The behavior of the masonry and the steel at different levels of load must be known to predict the strength of the member. Research suggests that there are four basic states in the behavior of a member undergoing flexural loading. They are:

1. *Uncracked section:* Strain levels in both the masonry and the steel are within the elastic limit. Further, the strain in the masonry is below that which is associated with cracking.
2. *Cracked section:* The strain levels in the masonry and the steel are still within the elastic range; however, the strain in the masonry exceeds the tensile limit. The material in the tension zone is not considered to contribute to the bending strength of the section.
3. *Inelastic behavior:* Strain levels in one of the components, masonry or steel, has exceeded its elastic limit.
4. *Ultimate strength:* The member reaches its ultimate capacity. The strain level in the masonry reaches ϵ_{mu}.

   Stages 1 and 2 correspond to the proportional part of the two
stress-strain curves shown in Figure 3.1. In Stage 1 the concrete masonry in both the tension and compression regions behaves elastically.
As the strains increase (Stage 2), the concrete masonry in the tension
region cracks and is assumed to carry no tension. The masonry in
the compression region still behaves elastically at this point. Stage 3
corresponds to that portion of the stress-strain curve for concrete
masonry above the proportional limit but less than the strain at
which the masonry fails in compression or when the strain level in
the steel exceeds the yield strain $\epsilon_s \geqslant \epsilon_y$. An elastic representation
of the behavior of the section is no longer possible. The fourth stage
occurs when the strain level in the concrete masonry reaches ultimate
strain, ϵ_{mu}.
   Current design philosophy favors the development of a reinforced concrete masonry member in which an explosive or sudden
brittle failure of the masonry in compression is avoided and yielding
of the tension steel limits the capacity of the member. The distress
of a masonry flexural member undergoing a tension failure is well
announced by the formation of tension cracks which occur before
the masonry in compression fails, often referred to as a *secondary
compression failure*. A force equilibrium approach is employed to develop the flexural theory which allows the designer to control the
proportioning of the masonry and steel so that this philosophy can
be put into practice.
   Figure 3.3 shows the force couple in a reinforced concrete masonry section with tension reinforcement in flexure. Flexural theory

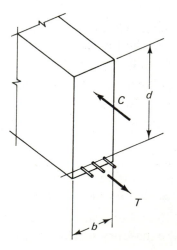

Figure 3.3 Force couple in a reinforced masonry member

recognizes that the internal compressive and tensile forces must remain in equilibrium. Therefore,

Tension force = Compression force

$$T = C \tag{3.11}$$

Equilibrium and the stress-strain relationship for the concrete masonry and steel make it possible to determine the location of the neutral axis and the magnitude of the tensile and compressive forces. This allows the designer to establish the capacity of the member.

At small levels of strain where the extreme fiber stress, f_m, is less than the tensile strength of the masonry, the section is uncracked and both stress and strain vary linearly from the neutral axis. The masonry in the tension and compression regions and the tension steel is considered effective in resisting the axial stresses induced by bending. In a manner similar to that developed in Section 3.2, the area of the steel can be transformed into an equivalent area of masonry using the modular ratio, n, as shown in Figure 3.4. The stress at the extreme compressive, f_{mc}, or tensile, f_{mt}, fiber of the section would then be determined by substituting the distance from the neutral axis to the extreme compression or tension fiber, c_c or c_t, into the expression

$$f_{mc} = \frac{Mc_c}{I_g} \qquad f_{mt} = \frac{Mc_t}{I_g} \tag{3.12}$$

where M = applied moment
I_g = moment of inertia of the gross uncracked section
c_c, c_t = distance from extreme compressive (tensile) fiber to the neutral axis

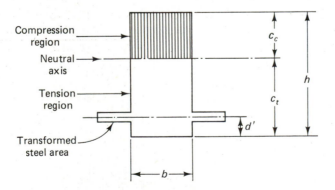

Figure 3.4 Uncracked transformed section

The stress in the steel is

$$f_s = n \left[\frac{M(c_t - d')}{I_g} \right] \tag{3.13}$$

where d' = distance from tensile face to centroid of steel

n = modular ratio = $\dfrac{E_s}{E_m}$

The limit of this range is defined by the cracking moment, M_{cr}, of the reinforced concrete masonry section:

$$M_{cr} = \left(\frac{f_{mt}}{c_t} \right) I_g \tag{3.14}$$

where f_{mt} is the tensile strength of concrete masonry.

Although this analysis is theoretically correct, it is seldom, if ever, used in actual design of a concrete masonry member because the tensile strain levels associated with cracking are exceeded in most members—if not as a result of bending stresses acting alone, then bending strains combined with shrinkage strains. The cracking moment is used primarily as an indicator of member serviceability to estimate or anticipate the behavior of a member when subjected to loads which will occur many times during the life of the structure. This is discussed in more detail in Chapter 4. The cracking moment is also used to establish minimum reinforcing requirements. (See Chapter 5.)

At greater levels of strain, small cracks in the masonry in the tension region begin propagating from the extreme fiber in toward the neutral axis as the tension strain is increased above the tensile strength of the masonry. The section is now considered "cracked" and only the steel is assumed to provide tension resistance. Figure 3.5a shows the linear stress distribution across a section in flexure where f_m is less than $0.5 f'_m$ (still in the elastic range) and the stress in the steel is less than f_y. The associated strain diagram is shown in Figure 3.5b. Figure 3.5c shows that the concept of the transformed area can still be used to establish the force/equilibrium relationship.

The location of the neutral axis, kd, can be determined by equating the moment of the transformed tension area to the moment of the compressive area about the neutral axis. An identical result can be obtained by equating the tension and compression forces in

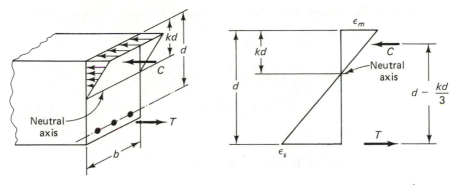

(a) Compressive Stress Block   (b) Strain Diagram

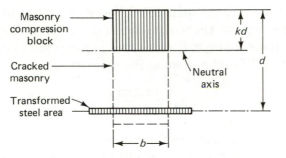

(c) Cracked Transformed Section

Figure 3.5 Cracked transformed section

the section. Referring to Figure 3.5, we have

$$T = C$$

$$\epsilon_m \left(\frac{d - kd}{kd}\right) E_s A_s = \frac{f_m}{2} bkd \qquad (3.15)$$

$$nA_s(d - kd) = \frac{1}{2} b(kd)^2$$

The unknown parameter k may be determined from Equation (3.15):

$$k = \sqrt{2\rho n + (\rho n)^2} - \rho n \qquad (3.15a)$$

where

$$\rho = \text{steel ratio} = \frac{A_s}{bd}$$

It can be shown that the expressions for the stress in the masonry and the steel are

$$f_m = \frac{2M}{bd^2jk} \tag{3.16a}$$

$$f_s = \frac{M}{A_s jd} \tag{3.16b}$$

where $j = 1 - k/3$

M = applied bending moment

If the masonry strain is still in the linear region when the steel strain reaches the yield point ($\epsilon_s = \epsilon_y$; see Figure 3.1), Equation (3.16b) can be used to calculate the moment at steel yield. The strain in the masonry can be checked to determine if it is still in the elastic range by assuming that $f_s = f_y$ and $\epsilon_s = \epsilon_y$. From the similar triangles shown in Figure 3.5b, the strain in the masonry is

$$\epsilon_m = \frac{\epsilon_y(kd)}{d - kd} \tag{3.17}$$

If, however, the masonry strain is not in the linear range, a non-linear model must be employed. Figure 3.6 shows the progression of the stress distribution in a flexural member through an increasing series of loads. As discussed above, at small strain levels [at (i)], the stress in the masonry varies linearly from the neutral axis, but with increasing moment the distribution becomes nonlinear. For a dis-

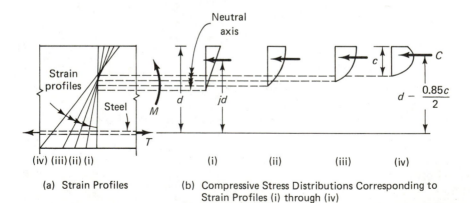

(a) Strain Profiles

(b) Compressive Stress Distributions Corresponding to Strain Profiles (i) through (iv)

Figure 3.6 Changing stress and strain distribution with increasing moment

cussion of how the compressive stress block changes in this transitional area in concrete, see Park and Paulay [3.5].

If, or as, additional load is applied to the member depicted by the stress distribution shown in Figure 3.6b the steel yields. As it yields, the strain in the steel and the concrete masonry increases. As the strain in the masonry increases, the compressive stress also increases, but since the compressive force, C, is limited by T_{max}, the neutral axis moves up as shown in Figure 3.6b [at (ii)]. The linearity assumed in Figure 3.6b [at (i)] soon becomes invalid and the depth of the compressive stress block continues to decrease until the masonry strain reaches its maximum value (ϵ_{mu}). This condition is depicted in Figure 3.6b [at (iv)] and represents the maximum or ultimate moment capacity of the member.

Tests of reinforced concrete members have shown that the exact shape of the stress block varies with respect to the cylinder strength and rate and duration of the loading. It is reasonable to expect that these same factors will similarly affect the precise shape of the compressive stress block in concrete masonry. The exact shape of the compressive stress block is not essential information if the designer can reasonably estimate the magnitude and location of the internal compression force resultant. This suggests that the actual flexural model may be simplified by replacing the true compressive stress distribution with an equivalent stress block. Once the designer locates the compressive force, the ultimate capacity of the member can be predicted.

The same approach has been used in the design of reinforced concrete, and the most commonly accepted equivalent stress block is one proposed by Whitney [3.6]. The Whitney stress block assumes a rectangular stress distribution. It seems reasonable to assume, in the absence of research in concrete masonry to the contrary, a similar equivalent rectangular stress distribution for reinforced concrete masonry members subject to flexure. Figure 3.7 shows Whitney's stress block with the assumed stress distribution superimposed upon it.

In a manner similar to that previously used for the linear stress case, the internal forces can be equated to determine the moment capacity. The magntude of the tension force T, is constant once the yield stress of the steel is reached, and must also equal the compression force C:

$$T = A_s f_y = C \tag{3.18}$$

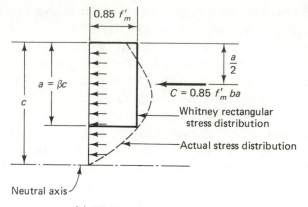

(a) Whitney Stress Block

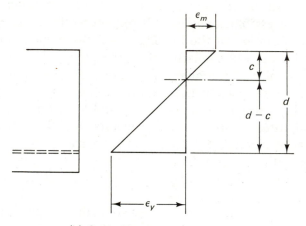

(b) Strain Diagram at Yield Load

Figure 3.7

The value of C can be determined by multiplying the magnitude of the stress by the area of the compressive stress block. Referring to Figure 3.7, we have

$$C = 0.85 f'_m \beta cb = 0.85 f'_m ab \qquad (3.19)$$

where $a = \beta c$.

The value of β has been established for concrete and it has been shown to be a function of f'_c. Little testing in concrete masonry has been done to determine the value of β and its relationship, if any, to f'_m. For values of f'_c below 4000 psi, β is equal to 0.85. It seems reasonable to assume that β is also equal to 0.85 for masonry; however, additional test data are recommended to increase our confidence in this value. Therefore, from Equations (3.18) and (3.19), and setting T equal to C, we have

$$A_s f_y = 0.85 f'_m a b \tag{3.20}$$

or alternatively,

$$a = \frac{A_s f_y}{0.85 f'_m b} \tag{3.21}$$

The nominal ultimate moment, M_n, is then given by

$$M_n = T \left(d - \frac{a}{2} \right) \tag{3.22}$$

or

$$M_n = A_s f_y \left(d - \frac{a}{2} \right) \tag{3.23}$$

Equation (3.23) can be rewritten using the value of C from Equation (3.19) in place of T:

$$M_n = 0.85 f'_m a b \left(d - \frac{a}{2} \right) \tag{3.24}$$

The assumption that the masonry strain in the compression region is in the inelastic range can be verified if it is assumed that $f_s = f_y$ and $\epsilon_s = \epsilon_y$. Using the similar triangles from Figure 3.7b gives us

$$\epsilon_m = \frac{\epsilon_y c}{d - c} \tag{3.25}$$

If the strain level in the masonry exceeds that associated with the ultimate strain, ϵ_{mu} (see Figure 3.1), prior to reaching a steel strain of ϵ_y, the masonry will fail prematurely or at subyield steel stresses, and Equations (3.22) and (3.24) no longer apply. This condition should be avoided and steel quantities limited to ensure that a ductile failure occurs. Chapter 5 presents a discussion of how ductile behavior can be achieved by limiting the maximum steel ratio, ρ_{max}.

When the depth of a member is restricted and sufficient masonry is not available in the compression region to carry the compressive stresses generated by the applied moment, or where tension is expected on both faces of the member because of load reversals

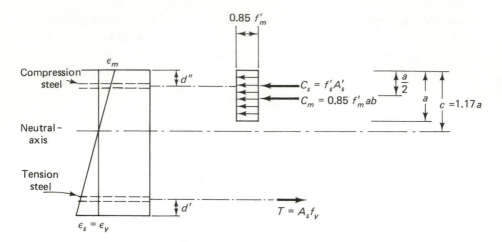

Figure 3.8 Flexural member with compression reinforcement

during lateral loading, steel can be added to the compression face. The flexural capacity of a member with compression steel can be developed by equating the internal forces in the same manner used in the preceding discussion.

Figure 3.8 shows the stress and strain distribution across a flexural member with compression steel. Instead of the compression region consisting of only the compressive force in the masonry, the steel in compression region also resists the imposed load. The total compressive force, C, can be expressed as

$$C = C_m + C_s \qquad (3.26)$$

where C_m = compressive force in the masonry
$\quad\ \ C_s$ = compressive force in the steel

The value of C is the sum of these two compressive forces:

$$C = 0.85f'_m ab + f'_s A'_s \qquad (3.27)$$

where A'_s = area of steel on the compression face
$\quad\ \ f'_s$ = compressive steel stress

and the value of T is

$$T = f_s A_s \qquad (3.28)$$

The nominal moment capacity of the double-reinforced member, M_n, is then given by

$$M_n = 0.85f'_m ab \left(d - \frac{a}{2}\right) + A'_s f'_s (d - d'') \qquad (3.29)$$

where d'' is the distance from compression face to the centroid of compression steel.

The strains in the masonry and the compression steel can be found as before using similar triangles. Designers frequently assume that both the tension and the compression steel have yielded at the ultimate moment because this assumption allows one to use f_y' in place of f_s' in Equation (3.29). This assumption can be verified as shown in Equation (3.30) by using the value of ϵ_m obtained from Equation (3.25):

$$\epsilon_s = \epsilon_m \left(\frac{c - d''}{c} \right) \geqslant \epsilon_y \qquad (3.30)$$

The ability to monitor the strain levels imposed on components of a beam and thereby avoid brittle failures is one of the principal advantages of ultimate-strength procedures. Ultimate-strength procedures also enable the designer to determine available member ductility. For example, if the strain level in the concrete masonry section shown in Figure 3.8 is ϵ_m when the steel first reaches yield ($\epsilon_s = \epsilon_y$), any additional deformation will produce a further straining of the steel and the concrete masonry. If the stress distribution in the concrete masonry at $\epsilon_s = \epsilon_y$ is already in the inelastic region of the stress-strain diagram the available section ductility may be closely approximated by taking the ratio of the ultimate concrete masonry strain, ϵ_{mu}, to the strain in the masonry, ϵ_m „ at the first yield of the steel. One definition of ductility, μ, is

$$\mu = \frac{\epsilon_{mu}}{\epsilon_m} \qquad (3.31)$$

An alternative definition requires specific information about the depth of the compressive stress block associated with variable levels of strain in the masonry. For a detailed discussion of this topic in concrete, the reader is referred to Park and Paulay [3.5].

Knowledge of available member ductility allows the designer to evaluate the ability of a concrete masonry member to undergo post-steel-yield load deformations. Volume 1 contains a broad discussion of the concepts of ductility.

EXAMPLE 3.4

Consider the reinforced masonry section shown in Figure E3.4-1. Assume that the stress-strain curves in Figure E3.1-1 apply. Assume that a moment, M, is applied about the X-X axis.

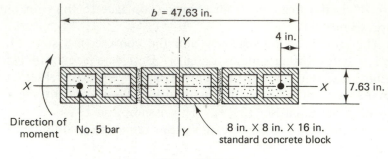

Figure E3.4-1

Using the information given above:

1. Determine the moment at which the masonry just cracks in tension if $f_{mt} = 0.135$ ksi.

2. Determine the moment, M_y, at which the steel yields, called the *yield moment.*

3. Assume a Whitney equivalent stress block and calculate the moment, M_n, at the ultimate strain, $\epsilon_{mu} = 0.0025$.

SOLUTION TO PART 1

Figure E3.4-2 shows the force/equilibrium relationship of the section. The neutral axis is located at the center of the wall because the stress and strain are equal on both faces. The steel *does not* contribute to the moment capacity at this level of strain.

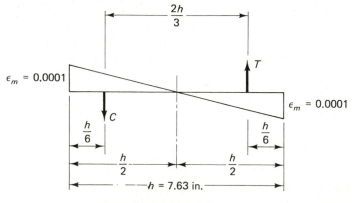

Figure E3.4-2

$$C = T = \frac{1}{2} f_{mt} \left(\frac{h}{2} \right) b$$

$$= 0.5(0.135 \text{ ksi}) \left(\frac{7.63 \text{ in.}}{2} \right) (47.63 \text{ in.}) = 12.27 \text{ kips}$$

$$M = T \left(\frac{2h}{3} \right)$$

$$= 12.27 \text{ kips} \left[\frac{2(7.63 \text{ in.})}{3} \right] = 62.39 \text{ in.-kips}$$

SOLUTION TO PART 2

The yield strain in the steel is $\epsilon_y = 0.00165$ and corresponds to a stress of 47.9 ksi. If we assume that a linear elastic stress-strain model is appropriate, and check to make sure that the masonry strain is still below the proportional limit ($\epsilon_m = 0.0005$), the modulus of elasticity of the masonry, E_m, is

$$E_m = \frac{f_m}{\epsilon_m} = \frac{0.703}{0.0005} = 1406 \text{ ksi}$$

Figure E3.4-3 shows the force equilibrium of the section.

$$T = C_m$$

$$= f_y A_s$$

$$= 47.9 \text{ ksi}(0.62 \text{ in.}^2) = 29.70 \text{ kips}$$

$$C_m = \frac{1}{2} (f_m) cb$$

$$= \frac{1}{2} (E_m) cb\epsilon_m \qquad \text{where } \epsilon_m = \epsilon_y \left(\frac{c}{d - c} \right)$$

$$= \frac{1}{2} \epsilon_y \left(\frac{c}{d - c} \right) E_m cb$$

$$= 0.5(0.00165) \left(\frac{c}{3.82 \text{ in.} - c} \right) (1406 \text{ ksi}) c(47.63 \text{ in.})$$

$$= 55.25 \left(\frac{c^2}{3.82 \text{ in.} - c} \right)$$

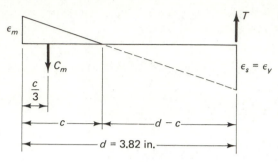

Figure E3.4-3

Solving for c yields

$$c = 1.181 \text{ in.}$$

The yield moment,

$$M_y = C_m \left(d - \frac{c}{3} \right)$$

$$= 29.70 \text{ kips} \left(3.82 \text{ in.} - \left(\frac{1.181 \text{ in.}}{3} \right) \right) = 101.8 \text{ in.-kips}$$

The value of the masonry strain is $\epsilon_m = 0.0007$, which is slightly higher than the estimated porportional limit of the concrete masonry; however, from the stress-strain curve it can be seen that the stress is still essentially proportional to the strain and a linear elastic stress-strain model is a reasonable approximation.

SOLUTION TO PART 3

$$a = \frac{A_s f_y}{(0.85 f_m' b)}$$

$$= \frac{(0.62 \text{ in.}^2)(47.9 \text{ ksi})}{(0.85)(1.5 \text{ ksi})(47.63 \text{ in.})} = 0.489 \text{ in.}$$

$$M_n = T \left(d - \frac{a}{2} \right)$$

$$= 29.70 \text{ kips} \left(3.82 \text{ in.} - \left(\frac{0.498 \text{ in.}}{2} \right) \right) = 106.1 \text{ in.-kips}$$

It can be seen that the difference between the yield moment, $M_y = 101.8$ in.-kips, and the ultimate moment, $M_n = 106.1$ in.-kips, is small (approximately 4 percent). On the other hand, the masonry strain for the ultimate moment at the outside fiber is approximately 3.5 times that at the yield moment, suggesting a considerable capacity to absorb energy even after the steel has yielded. The steel strain at $\epsilon_m = 0.0025$ is $\epsilon_s = 0.0171$ and is well within the elongation limits of a ductile steel such as Grade 40. This steel strain is approximately 10 times the yield strain and it is usually at this level of strain that reinforcing steels begin to strain-harden.

Figure E3.4-4 shows the moment/curvature relationship of the wall section in this example. The curvature, ϕ, of a section is

$$\phi = \frac{\epsilon_m + \epsilon_s}{d}$$

If *curvature ductility*, μ, is defined as

$$\mu = \frac{\phi_u}{\phi_y}$$

for the wall under discussion, the curvature ductility is

$$\mu = \frac{0.000615}{0.00513} = 8.34$$

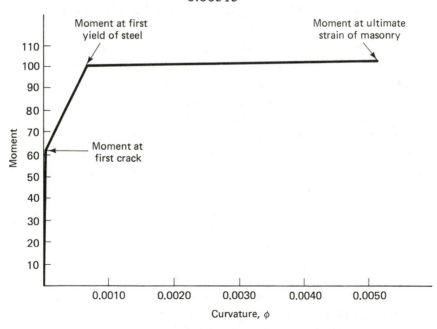

Figure E3.4-4 Moment/curvature relationship

This number gives some indication of the energy-absorbing capacity of the wall section, and examination of Figure E3.4-4 shows the amount of rotation that occurs from first yield of the steel to ultimate strain in the concrete caused by increasing the applied moment from 101.8 in.-kips to 106.1 in.-kips.

This example aids in the understanding of ductile reinforced masonry design and performance. Working stress design looks at structural performance from a load or force viewpoint. Therefore, a difference between 101.8 and 106.1 in-kip is small. However, the strength approach requires one to look *also* at deformations, and when this is done the ductile nature of masonry becomes evident.

3.4 MEMBERS SUBJECTED TO FLEXURAL AND AXIAL LOADS

Sections 3.2 and 3.3 developed expressions that describe the behavior of members subjected to axially or flexurally induced stresses. Although many elements in a masonry structure can be modeled as being subject to this type of loading, there are frequent cases where a member must be modeled as being loaded simultaneously in both flexure and axial compression or tension. This section presents a discussion and development of combined loadings in reinforced masonry members.

It is recalled from Section 3.3 that it requires several stages to describe the behavior of a member in flexure. The development used in that section can also be used to model the behavior of a member with combined axial and flexural loading.

An eccentrically applied compression load produces an equivalent concentric axial load of equal magnitude and associated moment. Figure 3.9 shows a member with an axial load, P, acting at a distance, e, from the centroid of the section. The moment, M, produced by this combination is given by

$$M = Pe \qquad (3.32)$$

If it is assumed that the section is uncracked, then any net tensile stresses must be less than the tensile strength of the masonry. This condition can arise in two ways: (1) The eccentricity of the axial load is so small that the tensile stresses produced by flexure are less than those required to cause cracking, or (2) the tensile stresses caused by flexure may be greater than the tensile strength of the masonry,

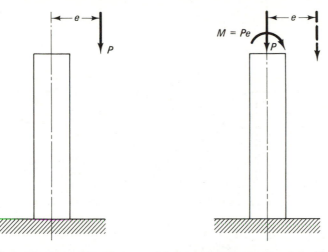

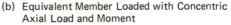

(a) Eccentrically Load Member

(b) Equivalent Member Loaded with Concentric Axial Load and Moment

Figure 3.9 Flexure and axial loads

but are reduced to the point where cracking does not occur by the applied compressive stresses. Figure 3.10 shows the superposition of stresses resulting from pure compression (f_a) and flexure (f_b) in a general masonry section with steel on both faces.

If the compressive stress is considered to have a *positive sign*, the stress caused by the concentric load, P, onto the uncracked transformed area, A_{ut}, is

$$f_a = \frac{P}{A_{ut}} \tag{3.33}$$

where A_{ut}, the area of uncracked transformed area, $= A_m + (n - 1)(A_s + A_s')$ and the axial stress associated with the moment produced by the eccentrically applied load is

$$f_b = \pm \frac{M(c_c \text{ or } c_t)}{I_{ut}} \tag{3.34}$$

where I_{ut} denotes the moment of inertia of the uncracked transformed section. The sign of f_b is both negative and positive because of the tension and compression stresses associated with flexure.

The combined elastic stresses on the uncracked section are now produced by adding the values obtained in Equations (3.33) and (3.34) as shown graphically in Figure 3.10.

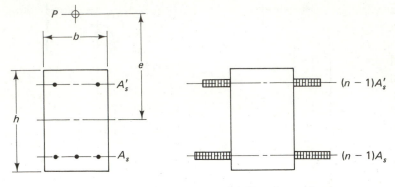

(a) Section under Combined Loading   (b) Transformed Section

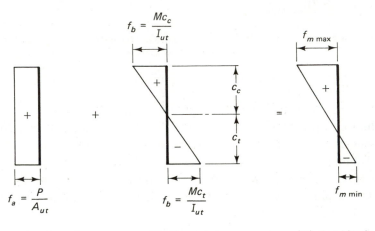

(i) Axial stress from P

(ii) Flexural stress from M

(iii) Combined stresses

(c) Combined Stresses

Figure 3.10 Uncracked eccentrically loaded member

$$f_{m\ \text{min}} = \frac{P}{A_{ut}} - \frac{Mc_t}{I_{ut}} \qquad (3.35a)$$

$$f_{m\ \text{max}} = \frac{P}{A_{ut}} + \frac{Mc_c}{I_{ut}} \qquad (3.35b)$$

If $Mc/I_{ut} < P/A_{ut}$, the entire section is in compression. This model is valid up to the point where the tensile strength of the masonry is exceeded. The values of P or e which represents the transition from a cracked to uncracked section can be found by setting

Equation (3.35a) equal to the tensile strength of the masonry, and replacing M with Pe.

As either the axial load or the eccentricity is increased beyond the point where the masonry tensile stress is equal to the tensile strength of the masonry, the section becomes cracked, and the masonry in the tension region is considered to be ineffective in resisting the applied loads. The expressions for the axial stresses must now consider the transformed *cracked* section as shown in Figure 3.11.

Using the principles of force equilibrium, the expression for the maximum stress in the masonry is

$$f_{m\,max} = \frac{P}{A_t} + \frac{Pec_m}{I_t} \qquad (3.36)$$

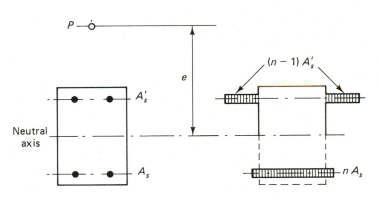

(a) Section Under Combined Loading   (b) Transformed Section

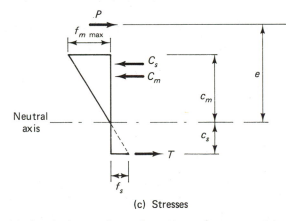

(c) Stresses

Figure 3.11 Cracked transformed section of an eccentrically loaded member

The tension stress in the steel is

$$f_s = \left(\frac{P}{A_t} - \frac{Pec_s}{I_t} \right) n \tag{3.37}$$

where A_t = area of transformed section
        I_t = moment of inertia of transformed section
        c_m, c_s = distance from the neutral axis to the outer fiber in the concrete masonry or the tension steel, respectively

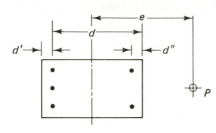

(a) Eccentrically Loaded Section

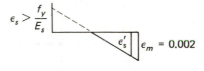

(b) Strain at Ultimate

(c) Stress from Flexure at Ultimate

(d) Stress from Compression

(e) Equivalent Stresses

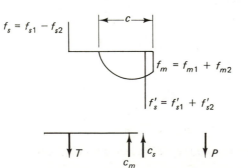

(f) Force Equilibrium

Figure 3.12 Eccentrically loaded member at ultimate load

Using an elastic model to describe the structural mechanics of a masonry member under a condition of combined loading is valid until the stress in the masonry exceeds the proportional limit (see Sections 3.2 and 3.3). At loads approaching the ultimate strength of the member, an inelastic redistribution of stresses occurs. Figure 3.12 shows an eccentrically loaded section at the ultimate load (again, a general masonry member with steel at both faces is shown) with the strain diagram.

As in a pure flexural member, failure can occur from either yielding of the steel or crushing of the masonry; however, unlike in a beam, limiting the tension steel will not always prevent a compression failure of the masonry because the type of failure is dependent on the level of the axial stresses. It is clear, then, that the usual assumption that the *tension steel* has yielded at the ultimate load is not necessarily correct. Whether or not the *compression steel* reaches the yield strength depends on the load level, yield strength of the steel, and the dimensions of the member. Consequently, there are three possible modes of failure: (1) yielding in the tension steel prior to the crushing of the masonry in compression (tension failure), (2) simultaneous yielding in the tension steel and crushing of the masonry (balanced failure), and (3) compressive crushing of the masonry prior to the yielding of the tension steel (compression failure). Figure 3.13 illustrates the strain diagrams at failure for various eccentrically loaded members. For Grade 60 steel, yielding of the compression reinforcing is not likely because $\epsilon_y \approx \epsilon_{mu}$, whereas for Grade 40 yielding is possible.

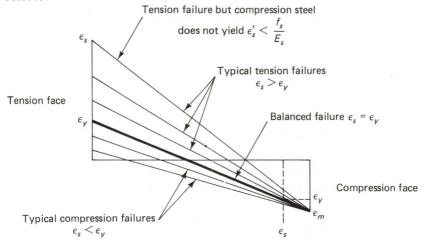

Figure 3.13 Strain diagram for eccentrically loaded member at failure

The general equation for the nominal compressive strength, P_n, is

$$P_n = 0.85 f'_m ab + A'_s f_{s2} - A_s f_{s1} \qquad (3.38)$$

The nominal flexural strength, M_n, may be found by summing moments about the centroid of the tension steel:

$$M_n = P_n e = 0.85 f'_m ab \left(d - \frac{a}{2}\right) + A'_s f_{s2}(d - d'') \qquad (3.39)$$

where a = depth of equivalent rectangular stress block
b = width of section
A'_s = area of compression steel

There are three undefined variables in this equation: f_{s1}, f_{s2}, and a. One solution procedure is to assume a strain distribution, thereby defining each variable. If, for example, points in the compression failure range are of interest, then $\epsilon_m = 0.002$ would define the strain on one face and various values of ϵ_s would be chosen. If ϵ_s is set at 0.001, f_{s1}, f_{s2}, and a become known and one can solve directly for P_n and M_n. Each value can then be plotted on an interaction diagram (see Section 5.3), thereby defining failure loads associated with these strains. Repeating this procedure for various values of ϵ_s, a failure envelope is then developed.

Where the entire cross section or a large portion of it is under compression, a is not a function of c (depth to the neutral axis) and other procedures must be used to define the compression force in the masonry. The stress-strain relationship of Figure 3.1 may be used or approximate modeling techniques developed as is done in Chapter 4.

EXAMPLE 3.5

Consider the pilaster shown in Figure 3.2 with material properties represented on the stress-strain curve shown on Figure E3.1-1. Assume that additional tests have been run and have determined that the tensile strength of the masonry is $f_t = 0.135$ ksi at a strain of 0.0001.

An axial load, P, is applied at an eccentricity of $e = 10$ in. about the Y-Y axis. Using this information, calculate:

1. The axial load, P, and moments, M, which will just cause the masonry on the tension face to crack.

2. The axial load, P, and moment, M, which will cause the ma-

sonry strain on the compression face to equal $\epsilon_{mu} = 0.0025$. Assume a Whitney stress block.

3. The axial load, P, and eccentricity, e, which will just cause the yielding of the tension steel and the masonry strain on the compression face to equal $\epsilon_{mu} = 0.0025$. Assume a Whitney stress block.

4. The moment capacity, M_n, of the pilaster (assume no axial load).

SOLUTION TO PART 1

The masonry on the tension face will crack when the strain equals $\epsilon_{mt} = 0.0001$ at a stress equal to 0.135 ksi. Summation of the vertical forces and the moments about any point in the section will provide the two equations to find the two unknowns, P and c. Figure E3.5-1 shows the force equilibrium relationships needed to solve for these two unknowns.

Summation of the vertical forces (neglecting the tension steel) gives an expression for P:

$$P = C_m + C_s - T_m$$

where

$$C_m = \frac{1}{2} \epsilon_m E_m bc$$

$$= \frac{1}{2} \left(\frac{\epsilon_{mt} c}{d - c} \right) E_m bc$$

$$= \frac{1}{2} \left(\frac{0.0001c}{15.63 \text{ in.} - c} \right) 1406 \text{ ksi}(15.63 \text{ in.})c$$

$$= \frac{1.01c^2}{15.63 \text{ in.} - c}$$

$$C_s = \epsilon_s' E_s A_s$$

$$= \frac{\epsilon_{mt}(c - d')}{(d - c)} E_s A_s$$

$$= \frac{0.0001(c - 2 \text{ in.})}{(15.63 - c)} (29{,}000 \text{ ksi})(0.61 \text{ in.}^2)$$

$$= \frac{1.77(c - 2 \text{ in.})}{15.63 - c}$$

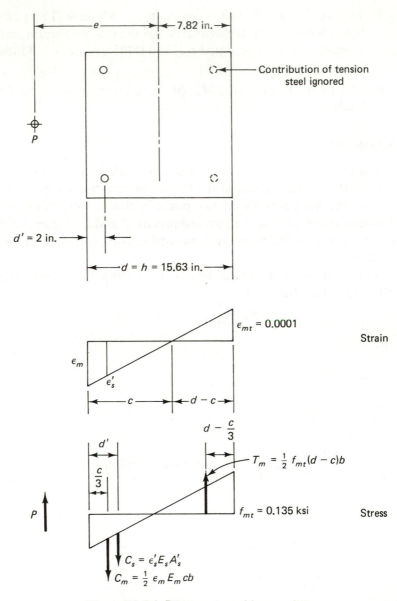

Figure E3.5-1 Pilaster at cracking condition

$$T_m = \frac{1}{2} \epsilon_m E_m b(d - c)$$

$$= \frac{1}{2} (0.0001)(1406 \text{ ksi})(15.63 \text{ in.})(15.63 - c)$$

$$= 1.10(15.63 - c)$$

Summing the moments about the tension resultant gives an expression to find P and c:

$$\sum M_{\text{at } T} = 0 = P\left[e + \frac{2(d - c)}{3}\right] - C_m \left[\frac{2c}{3} + \frac{2(d - c)}{3}\right]$$
$$- C_s \left[c - d' + \frac{2(d - c)}{3}\right]$$

Substituting into this equation the expressions for P, C_m, and C_s and then solving for P and c gives

$$c = 10.56 \text{ in.}$$

$$P = 20.63 \text{ kips}$$

The moment associated with an axial load of $P = 20.63$ kips at an eccentricity of $e = 10$ in. is

$$M = Pe = 20.63 \text{ kips } (10 \text{ in.}) = 206.33 \text{ in.-kips}$$

SOLUTION TO PART 2

Although the strain in the masonry has reached $\epsilon_{mu} = 0.0025$, it is not possible to know for sure whether or not the compression steel has yielded. It will be assumed that the compression steel has not yielded and once the compression steel strain has been computed, the assumption will be verified. Figure E3.5-2 shows the force equilibrium for this section.

Summation of the vertical forces gives

$$P = C_m + C_s - T_s$$

where

$$C_m = 0.85 f'_m ab$$
$$= 0.85 f'_m (0.85c)b$$
$$= 0.85(1.5 \text{ ksi})(0.85c)(15.63 \text{ in.})$$
$$= 16.94c$$

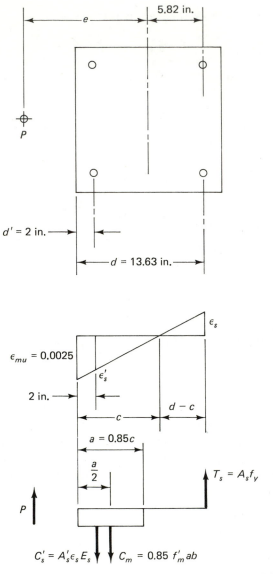

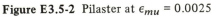

Figure E3.5-2 Pilaster at $\epsilon_{mu} = 0.0025$

$$C_s = \epsilon_s' E_s A_s'$$

$$= \frac{\epsilon_{mu}(c - d')}{c} E_s A_s'$$

$$= \frac{0.0025(c - 2 \text{ in.})}{c} (29{,}000 \text{ ksi})(0.62 \text{ in.}^2)$$

$$= \frac{44.95(c - 2 \text{ in.})}{c}$$

$$T_s = A_s f_y = 29.70 \text{ kips} \qquad (\text{where } \epsilon_s > \epsilon_y)$$

Summing moments about the tension resultant gives

$$\sum M_{\text{at}T} = 0 = P(e + 5.82 \text{ in.}) - C_m \left(d - \frac{a}{2} \right) - C_s(d - d')$$

Solving for c and P yields

$$c = 4.93 \text{ in.}$$

$$P = 80.53 \text{ kips}$$

The moment associated with an axial load $P = 80.53$ kips at an eccentricity of $e = 10$ in.

$$M = Pe = 80.53 \text{ kips} (10 \text{ in.}) = 805.26 \text{ in.-kips}$$

Verifying the assumption that $\epsilon_s' < \epsilon_y = 0.00165$ yields

$$\epsilon_s' = \frac{\epsilon_{mu}(c - d')}{c}$$

$$= \frac{0.0025(4.93 \text{ in.} - 2 \text{ in.})}{4.93 \text{ in.}}$$

$$= 0.00149 < 0.00165$$

It can be seen that the assumption was correct.

SOLUTION TO PART 3

At the balanced condition (steel yielding as the masonry crushes) the steel strain is $\epsilon_s = \epsilon_y = 0.00165$ and the masonry strain is $\epsilon_m = 0.0025$. From this information, the location of the neutral axis, c, can be determined from the similar triangles shown in Figure E3.5-3.

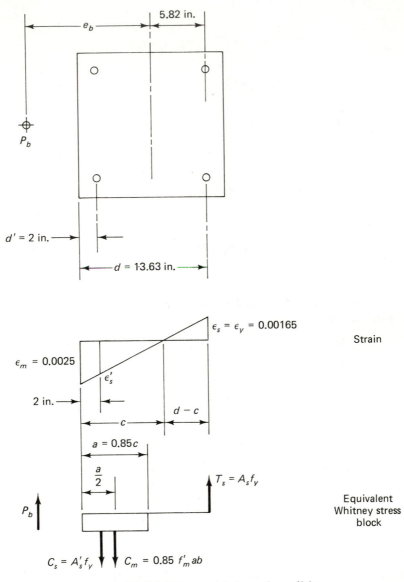

$\epsilon_s = \epsilon_y = 0.00165$

Strain

$\epsilon_m = 0.0025$

Equivalent
Whitney stress
block

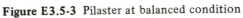

Figure E3.5-3 Pilaster at balanced condition

$$\frac{\epsilon_m}{c} = \frac{\epsilon_s}{d-c}$$

$$c = 8.21 \text{ in.}$$

$$a = 0.85c = 0.85(8.21 \text{ in.}) = 6.98 \text{ in.}$$

The strain in the compression steel, ϵ'_s, is

$$\epsilon'_s = \frac{\epsilon_m(c-d')}{c}$$

$$= \frac{0.00025(8.21 \text{ in.} - 2.0 \text{ in.})}{8.21 \text{ in.}}$$

$$= 0.0019 > \epsilon_y$$

It can be seen that the compression steel has yielded ($\epsilon' > \epsilon_y$), so from the force equilibrium shown in Figure E3.5-3, an expression for the axial load, P_b, may be found.

$$T_s + P_b = C_s + C_m$$

$$P_b = 0.85f'_m ab + A'_s f_y - A_s f_y$$

$$= 0.85(1.50 \text{ ksi})(6.98 \text{ in.})(15.63 \text{ in.}) + 0.62 \text{ in.}^2(47.9 \text{ ksi})$$

$$- 0.62 \text{ in.}^2(47.9 \text{ ksi})$$

$$= 139.10 \text{ kips}$$

The moment at the balanced condition, M_b, may be found by summing moments about the tension resultant and solving for the unknown eccentricity, e_b.

$$\sum M_{Ts} = 0 = P_b(e_b + 5.82 \text{ in.}) - C_m\left(d - \frac{a}{2}\right) - C_s(d - d')$$

$$= 139.1 \text{ kips }(e_b + 5.82 \text{ in.}) - 0.85(1.50 \text{ ksi})$$

$$\cdot (6.979 \text{ in.})(15.63 \text{ in.})\left(13.63 \text{ in.} - \frac{6.98 \text{ in.}}{2}\right)$$

$$- 0.62 \text{ in.}^2(47.9 \text{ ksi})(13.63 \text{ in.} - 2 \text{ in.})$$

$$e_b = 6.80 \text{ in.}$$

$$M_b = P_b e_b = 139.1 \text{ kips }(6.80 \text{ in.}) = 945.90 \text{ in.-kips}$$

SOLUTION TO PART 4

The moment capacity of the pilaster is that of a doubly reinforced beam. It is assumed that the compression steel has not yielded.

$$T_s = C_m + C_s$$

where

$$T_s = A_s f_y = 29.70 \text{ kips}$$

$$C_m = 0.85 f'_m ab$$

$$= 0.85(1.5 \text{ ksi})(0.85c)(15.63 \text{ in.})$$

$$= 16.94c$$

$$C_s = \epsilon'_s E_s A'_s$$

$$= \frac{\epsilon_{mu}(c - d')}{c} E_s A'_s,$$

$$= \frac{0.0025(c - 2 \text{ in.})}{c} (29{,}000 \text{ ksi})(0.62 \text{ in.}^2)$$

$$= \frac{44.95(c - 2 \text{ in.})}{c}$$

Solving for c yields

$$c = 1.90 \text{ in.}$$

It can be seen that "compression" steel is actually in tension because $c < d'$. Solving for the moment capacity gives us

$$a = 0.85c = 0.85(1.90 \text{ in.}) = 1.612 \text{ in.}$$

$$C_m = 32.13 \text{ kips}$$

$$C_s = -2.44 \text{ kips} \quad \text{(tension)}$$

$$M_n = C_m \left(d - \frac{a}{2}\right) + C_s(d - d')$$

$$= 32.13 \text{ kips} \left(13.63 \text{ in.} - \frac{1.612 \text{ in.}}{2}\right)$$

$$+ (-2.44 \text{ kips})(13.63 - 2 \text{ in.})$$

$$= 357.79 \text{ in.-kips}$$

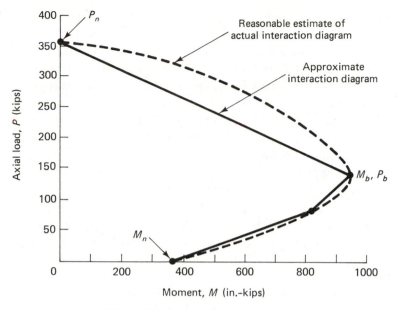

Figure E3.5-4 Interaction diagram

From the information gathered in this example and in Part 5 of Example 3.1, an interaction diagram may be constructed for this section. Figure E3.5-4 shows this interaction diagram and an estimate of the actual shape of the interaction diagram had more points been plotted. Section 5.5 discusses the construction of an interaction diagram in more detail.

3.5 MEMBERS SUBJECTED TO SHEAR LOADS

The failure of a reinforced masonry member in shear may be a very brittle failure producing an explosive crushing of the concrete masonry. The designer should strive to prevent these sudden brittle failures and allow the ultimate strength of the member to be controlled by the yielding of steel. There is, then, strong motivation to be able to predict accurately the ultimate strength of concrete masonry in shear.

Research work has been undertaken in concrete in an attempt to develop a rational theory to determine the shear strength of a member. Although these research efforts increase the understanding of the behavior of nonhomogeneous materials such as reinforced concrete masonry subject to shear, a complete understanding of shear-related phenomena has yet to be achieved. Many of the problems related to obtaining a rational approach to describe shear behavior are a result of the fact that concrete masonry is not an elastic, homogeneous material, a distinction that many other building materials share but one that is frequently ignored.

A masonry member is constructed of individual units and, therefore, may fail differently than a monolithic member. The influence of mortar joints and grouting on the shear strength of a masonry member are examples of conditions peculiar to masonry. On the other hand, masonry shear strength is affected by many factors common to concrete. As such, it appears reasonable to begin with a discussion of the mechanics of shear strength by examining the behavior of masonry as if it were a monolithic material and then examine the stress levels that develop along discontinuities such as mortar joints.

In order to predict the structural behavior on any plane in a member, an understanding of the stress patterns created in the member subjected to shearing stresses acting alone, or in combination with axial stresses, must be obtained. This will be the first task addressed in this section. The theory developed to support and understand the behavior of concrete members will be summarized, and equivalent relationships will be developed for concrete masonry, recognizing the stress limitations associated with concrete masonry construction.

Appropriate design procedures for reinforcing concrete masonry can be developed only if the engineer understands the implications and limitations of the theory and correlates them with available test data. This will be the objective of Chapter 5.

Planes of principal stress are a function of the ratio of shear stress to bending stress. For this reason, the ratio M/Vd or h/d becomes an important variable.

If a concrete beam in flexure is assumed to behave as a monolithic, homogeneous isotropic member, the principles of strength of materials show that the axial and shear stresses present in a differential element (Figure 3.14a) may be resolved into principal stresses oriented to the neutral axis of the beam at some angle α (Figure 3.14b). The magnitude of these principal stresses is given by

$$f_t = \frac{1}{2} f + \sqrt{\left(\frac{1}{2} f\right)^2 + v^2} \qquad (3.40a)$$

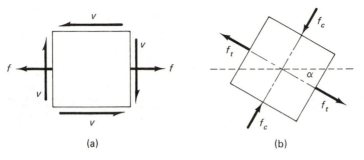

Figure 3.14 Orientation of principal stresses

$$f_c = \frac{1}{2} f - \sqrt{\left(\frac{1}{2} f\right)^2 + v^2} \qquad (3.40b)$$

where f = axial stress due to bending

v = shear intensity = V/bd

f_t, f_c = principal tension and compression stresses, respectively

The orientation of the principal stresses is given by

$$\alpha = \frac{1}{2} \arctan \left(\frac{2v}{f}\right) \qquad (3.41)$$

A differential element along the neutral axis has the principal stresses acting in tension and compression at an angle of 45° to the neutral axis. At elements other than on the neutral axis, however, the orientation of these stresses varies throughout the length of the beam and depends on the relative intensity of the component's flexural and shear stress. It can be seen from Figure 3.15 how the basic trajectory of these inclined forces is oriented for a uniformly loaded, simply supported beam.

The diagonal tension stress caused by the resolution of the shear and axial stresses is presumed to be responsible for the inclined cracking found in many flexural members. The principal tension stresses act on the concrete masonry, which is relatively weak in tension, and cause cracks perpendicular to their line of action. These cracks may develop both with and without the presence of flexural cracking. *Web-shear cracks* are also shear-caused inclined cracks which develop in an area previously uncracked by the flexure of the beam. *Flexure-shear cracks* are inclined extensions of vertical cracks formed previously.

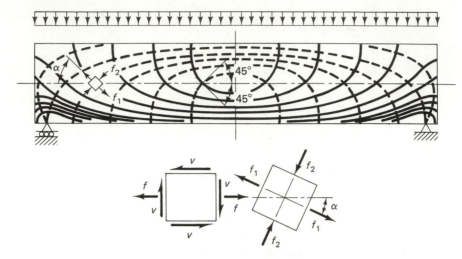

Figure 3.15 Stress trajectories [3.4]

In reinforced concrete beams without shear reinforcement, the transfer of the external force is resisted by several mechanisms (see Figure 3.16):

1. The shear force carried by the uncracked concrete in the compression zone

2. The vertical components of the inclined shearing stress similar to the frictional force caused by the interlocking of the aggregate particles along the cracked concrete

3. The dowel action of the flexural reinforcement resisting the transverse shear force

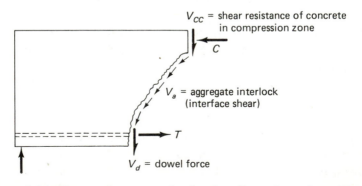

Figure 3.16 Shear resistance mechanism in a flexural member without web reinforcing

Of the three mechanisms, dowel action has been shown to be the smallest contributor to and most unreliable component of resisting shear loads in a beam without web reinforcement. The effectiveness of the dowel force is limited by the tensile strength of the concrete because the longitudinal cracking of the concrete parallel to the flexural steel reduces the stiffness of the section. Without the web reinforcement, the cracking of the concrete cover occurs at a lower level and severely reduces the effectivenss of the dowel action in resisting shear.

As a cracked section experiences shear deformation, the rough surfaces of the crack are brought into contact with one another. The projecting particles of coarse aggregate along the adjacent faces of the crack allow the transfer of many small shear forces. This aggregate interlock, or interface shear, mechanism is responsible for resisting a substantial amount of the imposed shear force.

The flexural cracking present in concrete beams under service loads is expected and causes no impairment to the beam's capacity to carry its design load. If, however, a member with no web reinforcement develops inclined shear cracks extending from the existing flexural cracks toward the compression region, the situation is quite different. Although it might appear that the only shear resistance available to the beam is that resulting from the uncracked compression zone, studies have shown that this region provides only 25 to 40 percent of the total shear resistance as the beam approaches failure. The remainder is still provided by the aggregate interlock and dowel force in the tension region [3.7]. As these remaining defenses against shear failure break down, the total shear force must then be carried across the compression zone, which is typically unable to arrest the crack propagation, and the beam experiences a diagonal tension failure. For this reserve capacity to be utilized, a redistribution of the shear stresses following the formation of the inclined crack must occur (aggregate interlock and dowel force). In light of the limited knowledge of this redistribution mechanism, however, the shear strength without web reinforcement in all but deep beams is assumed to have been reached with the formation of the first inclined cracks.

It is argued that the mechanics of shear in reinforced concrete are similar to those found in concrete masonry. It seems reasonable, then, to expect the same type of crack formation and propagation in members of either material. The fact that concrete masonry is not a homogeneous material introduces additional complexities into an exact explanation of the mechanics of shear. For example, the formation of the cracks in a concrete member is basically controlled by the tensile strength of the concrete, while the cracking pattern in concrete masonry is influenced by the existing weakened planes in the member formed by the mortar joints. However, behavior should still be similar between the two materials because of the similarity of the shear resistance mechanism supplied by the aggregate interlock in cracked concrete members and an equivalent action in the masonry. Shear tests on masonry piers [3.8] have characterized the progressive shear failures in concrete masonry. Figure 3.17 shows the cracking pattern for a pier subjected to increasing cyclic shear loads. The first cracks to form are along bed joints; then diagonal cracks progress along bed and head joints and shearing of blocks follows. This suggests two distinct shear mechanisms: a shear friction type of failure along the bed joints, and diagonal tension failure. If the shear friction failure is restrained by vertical reinforcing, diagonal tension cracks will form and the behavior of the pier should be quite similar to that of a concrete pier, and aggregate interlock should play an important role in the ultimate shear strength of the concrete masonry.

The exact nature of the growth and propagation of the inclined cracks in a reinforced member without web reinforcing is proportional to the ratio of the shear and flexural stresses and may be expressed as

$$\frac{f}{v} \propto \frac{M}{Vd} \tag{3.42}$$

The shear span, a, is the ratio of M to V, and for a simple beam with a concentrated load is the distance over which the shear is constant. If the shear is not constant, the shear span has a value at every point along the beam. Thus, the shear span-to-depth ratio is equivalent to the expression in Equation (3.42):

$$\frac{a}{d} = \frac{M}{Vd} \tag{3.43}$$

Most concrete masonry flexural elements are shear walls and the value of a is typically assumed to equal the height, h. The shear walls are typically modeled as cantilever beams and the development of the shear span-to-height equivalence arises from the fact that a cantilever beam resembles half of a simply supported beam with a concentrated load at midspan. The fact that many shear walls have loads

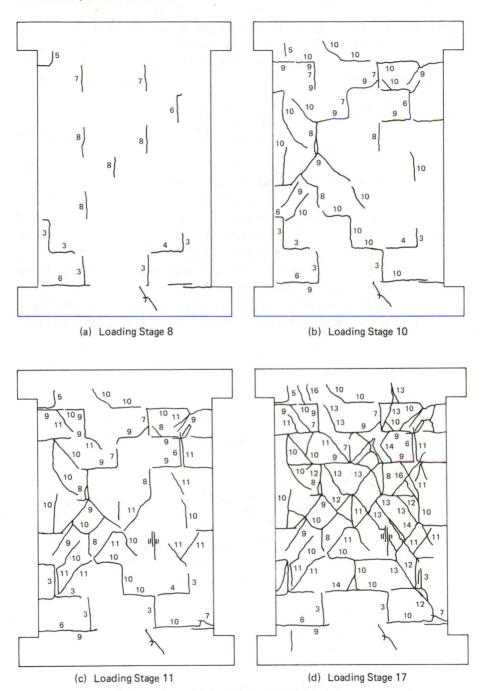

(a) Loading Stage 8

(b) Loading Stage 10

(c) Loading Stage 11

(d) Loading Stage 17

Figure 3.17 Crack propagation in wall 2 [3.8]

applied at several points up their height makes an exact equivalence of the h/d ratio to the a/d ratio somewhat inaccurate; however, the h/d ratio can still be used as a qualitative guide to shear wall behavior. Thus, the height-to-depth ratio is equivalent to the shear span-to-depth ratio:

$$\frac{h}{d} = \frac{a}{d} = \frac{M}{Vd} \tag{3.44}$$

The ratio a/d is an important variable, for it allows one to classify groups of flexural members based on their modes of failure. Figure 3.18, which plots the failure moment of a simply supported beam loaded with two symmetric point loads against the a/d ratio, may be used to identify the range over which each mode of failure occurs. The four basic categories are [3.10]:

1. *Long beams:* $a/d > 6$. The flexural capacity is less than the shear strength of the member. A failure of this type of member is exemplified by the yielding of the flexural steel and the eventual compressive failure of the concrete masonry.

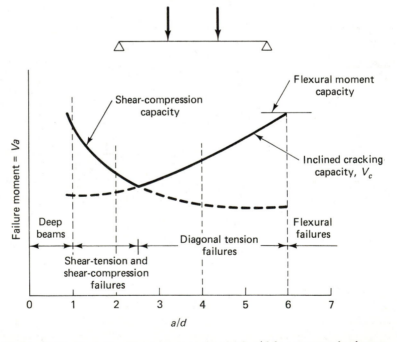

Figure 3.18 Variation in shear capacity with a/d for rectangular beams [3.9]

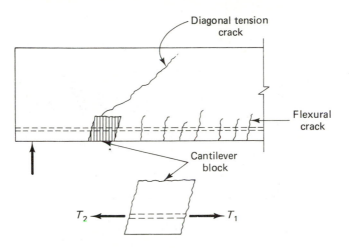

Figure 3.19 Diagonal tension failure in beams of intermediate length: $2.5 < a/d \leqslant 6$ [3.9]

Although short inclined cracks may form along the vertical flexural cracks, the magnitude of the shear force does not influence the ultimate strength of the beam.

2. *Beams of intermediate length:* $2.5 < a/d \leqslant 6$. In this type of beam the shear strength is equal to the inclined cracking strength. The cracks present from the flexural load cause numerous small blocks to form in the tension zone of the member, as shown in Figure 3.19. An increase of the tensile force in the flexural steel causes a bond force to form, $\Delta T = T_1 - T_2$ and, as the moment caused by this force increases, the moment at the base of the block increases. Increasing tensile stress in the steel causes additional flexural cracks to form, which reduce the width at the base of the blocks. If this width is reduced to the point where the stress caused by the moment is greater than the inclined cracking strength, an inclined crack forms and propagates up through the beam, causing it to fail in diagonal tension.

3. *Short beams:* $1 < a/d \leqslant 2.5$. Here the shear strength exceeds the inclined cracking capacity, and the load causing the inclined cracks is less than the load leading to failure. The inclined flexure shear crack (Figure 3.20a) extends into the compression zone with increasing load and grows downward to the tension reinforcement. A splitting crack usually de-

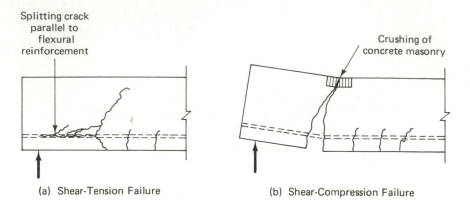

(a) Shear-Tension Failure   (b) Shear-Compression Failure

Figure 3.20 Shear failure in short beams: $1 < a/d \leqslant 2.5$ [3.9]

velops parallel to the flexural reinforcement. Two modes of failure are then possible. A shear compression failure results when the material near the compression face crushes (Figure 3.20b) or a shear-tension failure results from the flexural reinforcement failing at the anchorages (typically over the supports).

4. *Deep beams: $a/d \leqslant 1$.* The shear capacity exceeds the inclined cracking capacity in deep beams. A tied-arch mechanism (Figure 3.21) is formed when the inclined cracking occurs and the load is carried in tension by the flexural steel and by the "compression struts" formed by the inclined cracks. The failure modes for a deep beam are classified into four types:

1. Flexural steel anchorage failure at the reactions

2. Crushing failure at the supports

3. Crushing of the compression strut near the top of the arch, or the yielding of the tension steel (flexural failure)

4. Tension failure over the support or crushing of the concrete compression strut because of an eccentricity of the arch thrust

These four failure modes for a deep beam are shown in Figure 3.21.
The discussion presented above dealt with members without any web reinforcing. Most concrete masonry structures subjected to both shear and flexure are reinforced with some kind of web rein-

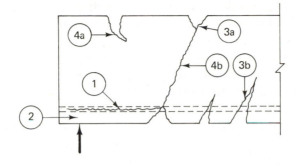

Legend

1. Anchorage failure

2. Bearing failure

3. Flexural failure
   a. Crushing of concrete
   b. Yielding of flexural steel

4. Arch-rib failure
   a. Over support
   b. At rib

Figure 3.21 Failure modes in deep beams: $a/d \leqslant 1$ [3.9]

forcing (i.e., the vertical and horizontal wall steel) and may exhibit behavior somewhat different from that described in the preceding discussion. The addition of web reinforcement, however, does not materially change the fundamental mechanics of resisting shear.

The addition of web reinforcement allows the designer to exert more control over the failure mode of the member. Web steel contributes to superior flexural performance (thus lessening the chances of an undesirable, uncontrolled diagonal shear failure) for several reasons:

1. It increases the effectiveness of the dowel action by tieing the longitudinal bars in place.

2. It restricts the growth of inclined cracks and improves the contribution of aggregate interlock.

3. It provides confinement pressure when the web reinforcing is spaced sufficiently close.

4. It improves the resistance to longitudinal splitting of the material and increases bond strength in the flexural steel.

5. It carries a portion of the shear force.

The most popular explanation of the shear strength in a member reinforced for shear is the truss analogy. It is explained by considering the action of the parallel chord truss shown in Figure 3.22a. The top and bottom chords act in compression and tension, respectively, and the web members act alternatingly in compression and tension. A reinforced beam with vertical shear steel (Figure 3.22b) may be thought of as resembling the truss shown in Figure 3.22a. The concrete, or concrete masonry, acting in compression forms the "top chord" and "diagonal compression struts." The vertical shear steel acts to resist the resulting vertical "tension" forces while the flexural tension steel forms the "bottom chord."

Considering the cracked flexural member with shear steel, shown in Figure 3.23a, an expression for the shear force carried by the steel, V_s, can be derived. The web reinforcement is inclined at

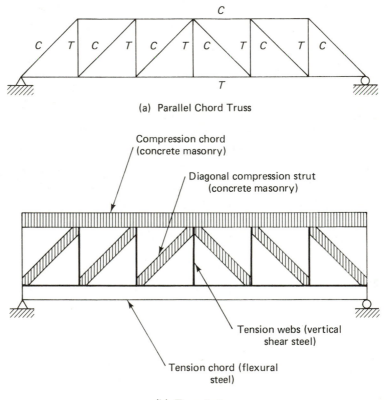

(a) Parallel Chord Truss

(b) Truss Action

Figure 3.22 Truss action in a reinforced beam with shear steel

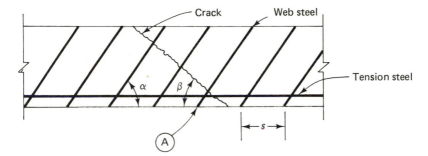

(a) Beam Elevation Showing Web Steel, Tension
    Steel, and Cracking

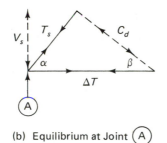

(b) Equilibrium at Joint (A)

Figure 3.23 Truss analogy

angle α and the compression strut is assumed to act at angle β. The
equilibrium conditions for joint A (Figure 3.23b) require that V_s be
equal to

$$V_s = C_d \sin \beta = T_s \sin \alpha \qquad (3.45)$$

where T_s is the resultant force from all web steel crossing the crack.

If it is assumed that the resisting moment of the beam has an
internal moment arm approximately equal to d, the stirrup force per
unit length of the beam, T_s, is

$$T_s = \frac{V_s}{d \sin \alpha(\cot \beta + \cot \alpha)} \qquad (3.46a)$$

$$= \frac{A_v f_s}{s} \qquad (3.46b)$$

where s = spacing between web reinforcement

A_v = area of shear steel spaced at s along beam

f_s = shear steel stress

Solving for A_v from Equations (3.46a) and (3.46b) and assuming that $f_s = f_y$ yields

$$A_v = \frac{sV_s}{df_y \sin \alpha (\cot \beta + \cot \alpha)} \tag{3.47}$$

Most web steel in concrete masonry is oriented vertically or horizontally. Thus, $\alpha = 90°$. It is commonly assumed that the orientation of β is at $45°$, and A_v is, from Equation (3.47),

$$A_v = \frac{sV_s}{df_y} \tag{3.48}$$

The web steel is not stressed until the member cracks. The shear carried by the masonry at the time the crack forms is V_m, and any additional load is assumed to be carried by the shear reinforcement. Thus, the total nominal ultimate shear strength of the beam, V_n, is the sum of the shear carried by the concrete masonry and that carried by the steel:

$$V_n = V_m + V_s \tag{3.49}$$

Special design considerations related to the interaction of V_s and V_m, as well as the minimum and maximum amounts of web steel, are discussed in Chapter 5.

In view of the fact that it is the diagonal tension failures in beams without web reinforcing that are so undesirable, the addition of axial compression to a member would be expected to increase the shear capacity of a member. Such an increase has been found to exist in concrete members [3.11]. It would appear reasonable to expect similar behavior in concrete masonry, and the limited data developed [3.8] support the contention that shear strength does, in fact, increase with the addition of a significant amount of axial load.

Although there are insufficient test data discussing the effects of axial tension on concrete masonry members subjected to shear, some observations gained from tests on concrete members may be instructive. It was found in one study [3.12] that the shear capacity of a reinforced concrete member subjected to axial tension was approximately 30 percent above that predicted by the appropriate design equations. The investigators concluded that tension reduced the shear capacity of the aggregate interlock, uncracked concrete, and dowel action mechanisms in resisting shear; but did not affect the performance of the truss action. This, they felt, accounted for the somewhat unexpectedly high shear capacity exhibited by these beams.

Most of the preceding discussion has treated the behavior of concrete masonry subjected to shear as if it were a monolithic material. It is not, of course, and one would expect that the existence of the mortar joints must have some influence on the behavior of a concrete masonry member. These planes of weakness usually exhibit cracking because of external tension, shrinkage, or other causes prior to the imposition of any shear stresses on the member. Premature failure along these weakened planes must be prevented so that aggregate interlock and dowel action may develop. Shear strength along these lines can only be developed by shear friction [3.13]; this suggests that horizontal and vertical reinforcement are essential to developing the shear strength of a concrete masonry section. Shear friction assumes that shear across the mortar joint is transferred by truss action (Figure 3.23). The transverse steel develops a clamping force across the cracked plane only after the crack appears. The transfer of shear across a mortar joint requires a relatively large shear displacement before the mechanism along the joint can form.

3.6 DEEP BEAMS

Walls are traditionally modeled and designed as elements in direct bearing. Most masonry flexural elements have relatively large span-to-depth ratios and are typically modeled and designed using traditional flexural theories. Walls that are not supported on continuous footings but span between isolated pad footings no longer carry loads in direct bearing but in flexure, and their span-to-depth ratios are generally much smaller than those of typical masonry beams. Shear walls often fall into this category, as typically their span-to-depth ratios are also small. A discussion of the structural mechanics of these deep beams is therefore required.

Beams with a small span-to-depth ratio may be analyzed using the concept of equating the compression and tension forces and establishing the magnitude of the internal moment arm to estimate their ultimate flexural strength. Span-to-depth ratios used in this section refer to panel length divided by panel height (l/h). This parameter is traditional in the literature of deep beam theory and is an important variable, as it characterizes shear flow, shear lag, and arch action in the beam. This is essentially the same phenomenon as that presented in Section 3.5, where it is discussed using the concept of shear span ($a = M/V$). In deep beam theory, a distribution is made to relate shear span and length parameters by treating simple span beams and continuous beams as separate entities. The location of the compression resultant will change with different span-to-depth ratios, as will the magnitude of the internal moment arm. Walls designed as deep beams should still be reinforced with the required minimum wall steel, but special attention must be paid to unusual conditions in these members, such as tied-arch action, anchorage requirements for the reinforcement, and the increase in diagonal tension and compression stresses near the supports at the onset of cracking.

There are two general classes of deep beams to be discussed: (1) moderately deep beams ($2 \leqslant l/h \leqslant 5$), and (2) very deep beams ($l/h < 2$); see Figure 3.24. The moderately deep beams are similar to beams of common proportions in most respects. Very deep beams will require a special discussion of the stress distribution.

It has been shown that for moderately deep beams with a span-to-depth ratio of $2 \leqslant l/h < 5$, regular methods of analysis appear to predict accurately the flexural strength of the member despite the fact that these sections carry a substantial portion of their load through a tied-arch mechanism. The value of modeling a deep beam as a tied arch is apparent when failures of deep beams at loads well below their computed flexural capacity are examined.

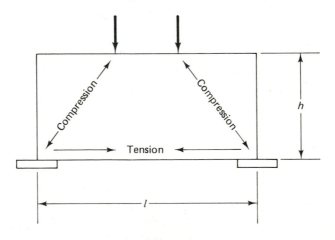

Figure 3.24

Figure 3.24 shows the load path in a deep beam modeled as a tied arch. The tension force at the bottom edge of the wall is essentially constant along the entire length of the span. Consequently, the flexural steel requirement established from a traditional method of analysis should be carried across the entire length of the beam because it functions as a tension tie between the two supports. Failures of deep beams at loads much smaller than the calculated strength have been reported [3.14] and were caused by inadequate anchorage of the tension bars at the supports. The magnitude of the tension force in the bars at the supports has been estimated at nearly 80 percent of the maximum tension force [3.14]. The ability of a member to reach its calculated flexural strength is dependent on preventing premature failures related to inadequate reinforcing anchorage.

The distribution of the tensile steel is generally along the tension edge, but because of the relatively large tension zone, it is recommended that the bars be distributed over a vertical distance equal to $0.25h - 0.05l$, where $h \leqslant l$ [3.5]. This distribution may be accomplished through the use of relatively small bars. The use of small bars also helps increase the effective development length and reduce anchorage problems because of the larger aggregate surface area available in the smaller-diameter reinforcing bar.

The tied-arch mechanism is most effective in beams with a shear span-to-depth ratio less than $2(M/Vd < 2)$. In beams with a uniform load, the shear span or span-to-depth parameter becomes the M/Vd ratio. The tied-arch mechanism should preferably be loaded from the top edge of deep beams. Beams that have loads hung from the bottom edge behave somewhat differently; however, since most masonry walls designed as deep beams are loaded along their top edge, the discussion that follows will be limited to this category.

Whereas the flexural capacity of a moderately deep beam and shear-induced stresses because the steel contributes to shear strength theory, the distribution and orientation of shear and diagonal tension stresses are significantly different from that found in a beam of common proportions. The shear strength of a deep beam may be two to three times that predicted by conventional theories [3.4]. A substantial portion of this increased strength may be explained through the tied-arch mechanism of load transfer.

It can be seen from Figure 3.25a that inclined tension cracks form parallel to the assumed load path between the applied vertical load and the reaction at the support. These cracks close to the sup-

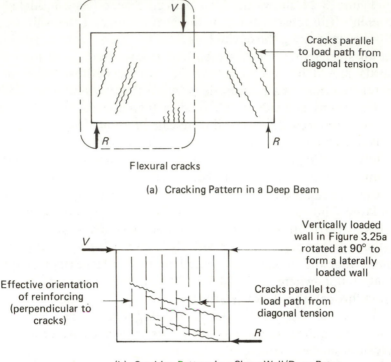

(a) Cracking Pattern in a Deep Beam

(b) Cracking Pattern in a Shear Wall/Deep Beam
(Rotated 90° from Figure 3.25a)

Figure 3.25 Cracking patterns in deep beams

ports are oriented at a much steeper angle than those found in traditional beams.   Inasmuch as the primary weapon against excessive
shear cracking is the rebar that crosses the cracks, the near-verticality
of the cracks in a deep beam suggests that a vertical orientation of
shear reinforcing still common in most beams is not particularly
effective.   The horizontal steel is more effective in resisting these
shear-induced stresses because the steel contributes to shear strength
through shear friction by acting at a direction approximately perpendicular to the crack.

Shear walls with low h/d ratios behave in much the same fashion
as deep beams.   As the tied-arch mechanism begins to form, tension
cracks form parallel to the applied horizontal load path as shown in
Figure 3.25b. Reinforcing, if it is to increase the capacity of the wall,
must be placed normal to the cracks.   The most effective orientation
of the shear steel in a laterally loaded shear wall with a low h/d ratio

would be a vertical orientation because the cracking essentially occurs in the horizontal plane.

The ACI Code [3.15] has developed a method to predict the ultimate shear capacity of a deep beam which presumes that the shear capacity is given by the superposition of the shear strength of the concrete and that of the web steel. In addition to the foregoing contributions to the shear capacity that are assumed in conventional beam shear, the ACI considered the reserve shear capacity of deeper beams (discussed in Section 3.5) in developing their model. Unfortunately, no comparable research has been done in concrete block masonry; however, test data clearly indicate a considerable increase in shear strength in members with low h/d ratios [3.16].

It appears that current methods of analysis will have to be used until a better explanation of the shear capacity of deep masonry members becomes available. Analyses made using existing tools appear to underestimate the ultimate shear strength by a significant, but at this point unquantifiable amount. Shear steel should, however, be provided in both the vertical and horizontal direction.

For deep beams with span-to-depth ratios smaller than 2.0 for a simply supported beam and 2.5 for a continuous beam, the flexural capacity cannot be predicted by conventional bending theory. A fundamental assumption in the development of traditional flexural theory is that plane sections remain plane (i.e., the strain distribution is linear). Deep beams exhibit deviations from this assumption because of a significant shear warping of the cross section. The concept of strain compatability still applies to deep beams, but traditional methods can no longer be applied to establish the section properties of the member or the stress distribution from the neutral axis.

Figure 3.26 shows the experimentally determined distribution of flexural stresses for a simply supported beam or wall for several span-to-depth ratios. For larger span-to-depth ratios the stress distribution is close to that assumed in traditional flexural theories. As the section depth increases with respect to the span length, the deviation from the typical stress distribution becomes more pronounced and the location of the compression resultant moves closer to the tension resultant. If the internal moment arm is given by z, it is seen from Figure 3.26 that it does not change significantly for different span-to-depth ratios.

The approximate value of the internal lever arm, z, has been established for various span-to-depth ratios for simply supported

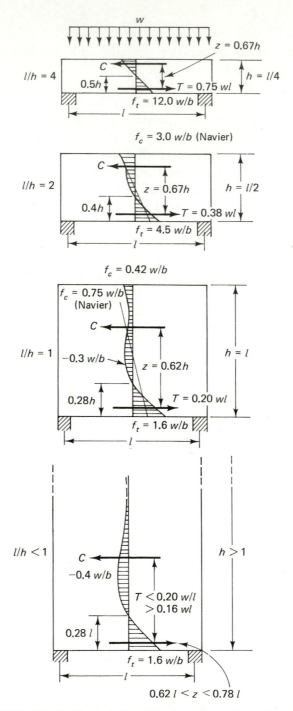

Figure 3.26 Distribution of flexural stresses in a homogeneous simply supported beam [3.5, 3.17]

beams [3.14]. For purposes of analysis, z may be assumed to be equal to the value determined by

$$z = 0.2(l + 2h) \quad \text{if } 1 \leqslant l/h \leqslant 2 \quad (3.50a)$$

or

$$z = 0.6l \quad \text{if } l/h < 1 \quad (3.50b)$$

In the equations above, the value of l is the smaller of 1.15 times the clear span or the center-to-center distance between the supports.

Deep-wall beams, spanning over several isolated footings, are basically continuous deep beams. In these members, the stresses and strains are not proportional even at very low strains. As the span-to-depth ratio approaches 1, the magnitude of the moment arm between the internal tension and compression forces decreases quite rapidly. As a result of this decrease in magnitude of the lever arm, it is possible that the tension force over the support (negative moment region) is closer to the compression face than it is to the tension face. Figure 3.27 shows the distribution of stress in a continuous beam for various span-to-depth ratios.

In a manner similar to that used to determine the magnitude of the internal lever arm for positive and negative moments for a simply-supported deep beam, z is given by the following expressions [3.14] for continuous deep beams:

$$z = 0.2(l + 1.5h) \quad \text{if } 1 \leqslant l/h \leqslant 2.5 \quad (3.51a)$$

or

$$z = 0.5l \quad \text{if } l/h < 1 \quad (3.51b)$$

The bending moments in a continuous deep beam are estimated as if the beam were a beam of normal proportions. Thus, for a beam with a uniform load, the moments are $wl^2/24$ at midspan and $wl^2/12$ at the support.

The value of z depends on whether the section under analysis is at midspan or over the supports; z will likely be smaller over the supports than at midspan. Equation (3.51) does not account for this change. It has been shown, however, that the actual moment over the support is approximately 15 percent smaller than $wl^2/12$ if the bearing area of the wall on the footing is 10 percent of the clear span between the footings. Consequently, the smaller value of z expected over the support is compensated for in an indirect manner.

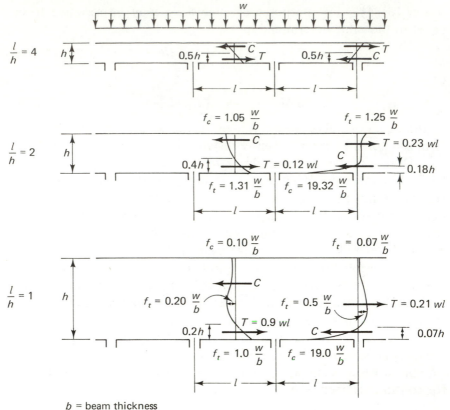

Figure 3.27 Distribution of flexural stresses in a homogeneous continuous beam [3.18]

The analysis of a deep beam with concentrated loads could be accomplished in a similar manner with the moments determined as if the beam were a slender member. The value of z is somewhat different, but [considering the empirical nature of Equation (3.51)] not so different that satisfactory results cannot be obtained.

The previous discussion addresses only two or three simple beam geometries and boundary and loading conditions. Deep beams exhibiting physical configurations different from the examples above require more refined analytical and experimental techniques to predict their behavior.

4

MODELING AND DESIGN CONSIDERATIONS

4.1 GENERAL

All too often the design engineer is submerged in the morass of complex equations that structural engineering appears to have become, and tends to lose sight of the physical significance of the problem that is being solved. The good engineer will understand what is happening to the structure under design when it is subjected to the environmental activity it must resist, and will use this understanding of building behavior to monitor the conclusions suggested by analysis.

How do engineers develop an understanding or feeling for structures? "Experience" is the standard answer—end of discussion. That certainly is an important part of the answer, but it is not the whole story. Let us examine a typical task and see if it sheds some light on the subject. Given the task to determine the deflection of a beam, the mathematician must consider all contributors to deflection: bending, shear, and so on. The engineer, repeating this process a few times, quickly sees that bending stiffness accounts for most of the deformation and shortcuts the design process by neglecting shear and the other contributors to deformation. The engineer has created an analytical model, and now can predict the force/deformation relationship in a member because it is understood how characteristics of the system affect its behavior. A feeling for how the beam will behave has been developed. This "modeling" process obviously expedites the design and will produce acceptable designs, as long as the assumptions used to eliminate the extraneous variables remain valid. Consciously or unconsciously, the engineer uses analytical models in almost all phases of the design process. In this chapter the thought process and resulting assumptions used in the development of analytical models are discussed.

The modeling process can be subdivided into two areas. First, the environmental impact on a structure must be reduced to a load or force criterion for the design of the components of the structure; this category of modeling is discussed in Sections 4.2 through 4.6. Once the load or force criterion for a component has been established, the adequacy of the component must be investigated. Usually, this involves converting forces to stresses. The appropriate component model must be developed. This type of modeling is discussed in Sections 4.7 through 4.10.

4.2 MODELING THE SYSTEM

The engineer most often thinks of the system model as one that helps to define the path a load will travel through a structure. To determine the distribution of forces along a shear wall with many openings or allocate forces between walls and frames, the engineer must develop an analytical model of the components of the system. A "pure" solution to this type of problem is impossible, as the variables involved are too numerous and complex. Most of these variables must be eliminated, or their complexity significantly reduced, if system behavior is to be understood.

The most important consideration is the significance of the contribution of a variable element to the behavior of the system. The force distribution to participating components is typically accomplished assuming that elastic stiffness is sufficiently representative of component behavior. Often, elastic behavior is not only inappropriate, but complicates the modeling procedure unnecessarily. In this section some of these system modeling decisions are discussed.

Load quantification is another extremely important aspect of system modeling when earthquake loads are a factor. During an earthquake, a building vibrates and the dynamic properties of the system must be understood because the force level is a function of these properties.

For example, an extremely rigid structure will tend to behave as though it is linked to the ground, and hence has the same behavior pattern as the ground on which it rests. As the structure becomes more flexible, it tends to resonate during an earthquake and amplifies the ground motions. A further softening of the structure tends to isolate the structure from the ground motion. System modeling has a significant impact on earthquake load development. Some of the considerations dealing with the significance of component modeling on earthquake load prediction are discussed in Volume 1. They will be discussed in more detail herein as they apply to various types of low-rise concrete masonry buildings.

Two features are common to all buildings, mass and stiffness (or bracing). The distribution of mass and stiffness must be understood if the designer is to develop components with appropriate levels of strength. Let us explore modeling techniques by examining a simple building system. The system shown in Figure 4.1 has both mass and stiffness. The mass is inherent in all of the building components. The stiffness characteristics are provided by the walls and the diaphragm. If we examine the behavior of an element of wall A (see Figure 4.1a) during an earthquake, we can develop an understanding of the load path and how to model the system. Let us assume that the building is rigid enough to move with the same motion as the ground. The wall is then subjected to an acceleration and this imposes an inertial force on the wall equal to the product of its mass times acceleration. This is commonly modeled as an equivalent static load; in other words, the designer views this load as being no different from a constant or static pressure acting on the wall. The wall must now span from the ground to the roof and be capable of sustaining these earthquake-induced inertial loads (Figure 4.1b). The wall inertial loads must now be transferred to the vertical bracing system by the horizontal bracing system, the diaphragm (Figure 4.1c). As the inertial loads travel through the diaphragm, they are joined by inertial loads which are produced as a result of the mass of the diaphragm undergoing an acceleration. Upon reaching the vertical bracing system, these inertial loads flow to the ground through the shear wall (Figure 4.1d). Once again, they are

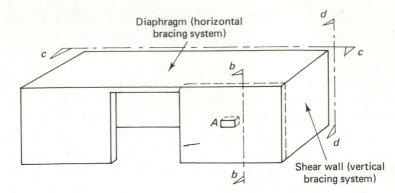

(a) Building Model for Lateral Loads

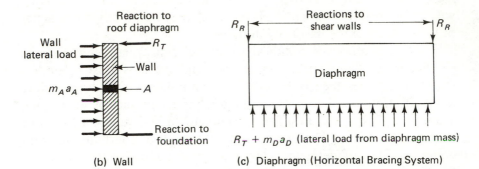

(b) Wall

(c) Diaphragm (Horizontal Bracing System)

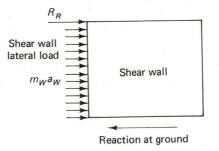

(d) Shear Wall (Vertical Bracing System)

Figure 4.1 Lateral load flow

joined by inertial loads produced in the wall by the acceleration acting on the mass of the bracing itself. The complex dynamic behavior of this simple system is modeled by:

1. Assuming an acceleration to which the system and its components will be subjected to during an earthquake. This is

accomplished in practice by replacing inertia forces with a set of equivalent static lateral forces.

2. Allocating the appropriate amount of mass supported by each bracing system.

3. Developing the appropriate model for each component of the lateral bracing or force-resisting system.

The acceleration, or more precisely the design acceleration, is estimated in a variety of ways. Code procedures prescribe component load levels: sometimes in the form of a percentage of base shear, and other times as a percentage of component weight. In either case a design acceleration in terms of a percent of gravity (*g*) is developed. Alternatively, spectral procedures may be used to develop a design acceleration level. Spectral procedures require the designer to develop a model for the system. A discussion of predicting design accelerations is presented in Volume 1.

4.3 MODELING THE WALL FOR OUT-OF-PLANE LOADS

For out-of-plane loads the wall serves two functions (see Figure 4.2). It must span vertically between horizontal bracing elements (floor or roof diaphragms) and it must transmit vertical loads to the foundation. Three conditions must be dealt with in modeling the wall:

1. How is the mass proportioned?
2. How does the lateral load travel to the supporting diaphragms?
3. What is the appropriate model for establishing stability criteria?

The *distribution of mass* to bracing elements, although important, should not occupy too much of the designer's time. Most designers assume that the parapet plus half the story height must be supported by the roof diaphragm (Figure 4.2) and half of the story height above and half of the story height below a floor is supported by the floor diaphragm. Occasionally, more sophisticated models are used by designers or required by reviewers but this is usually not warranted, as the other variables in the design are not defined to this level of accuracy.

Consider now the *load path*. Walls are considered to behave as

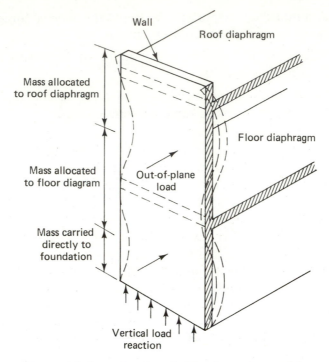

Figure 4.2 Lateral and vertical load-carrying model

one-way slabs in that they carry an out-of-plane lateral load in the shorter of two orthogonal directions. Thus, if the floor-to-floor (vertical) height is less than the distance between pilasters, the wall is assumed to span vertically. The span would be measured horizontally if the pilasters are placed closer together than the vertical supports. If the vertical height between supports and the horizontal distance between pilasters is nearly equal, it would seem reasonable to approach the modeling of the wall as a two-way slab; however, there currently is little research assessing the performance of walls designed in this manner or research on the influence of the boundary elements (foundations, floor slabs, pilasters, and the like) on the wall.

If a wall of height x, spanning vertically, can develop fixity at its base and is pinned at the top, it may be modeled as a propped cantilever beam. A point of inflection may be reasonably assumed to exist at a point $0.2x$ from the fixed support, a point considered to be a brace point for stability considerations. Thus, the unsupported height h is now reduced to $0.8x$.

Flexural resistance at the base is provided by the moment couple produced by the slab-wall dowels and the force resulting from the friction force at the bottom of the wall footing. Figure 4.3 shows this force couple. The magnitude of the moment is given by the applied uniform lateral load over an assumed cantilever beam of height $0.2x$ and the shear at the end of the cantilever, R, resulting from the applied uniform load over the remaining $0.8x$ of the wall. The wall itself is often designed for the moment produced by the uniform load acting over the entire height of the wall, including the lower portion of the wall as though it were a simply supported beam. This is conservative and the actual moment diagram may be used if the design effort is warranted. Figure 4.4 shows examples of vertical wall fixity conditions.

In a multistory wall, the moment of a vertically spanning wall is typically found by assuming that the continuous wall passes over supports at each floor. For stability considerations, however, this is not appropriate unless the wall is restrained by the floor. If adequate restraint is provided, the effective height is reduced to approximately $0.6x$ or $0.65x$.

Connection between the wall and the floor system must be maintained during the lateral motions. If the floor or roof tie connection fails, as has happened during earthquakes, it removes a brace

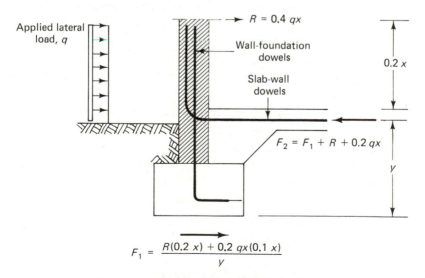

$$F_1 = \frac{R(0.2\,x) + 0.2\,qx(0.1\,x)}{y}$$

Figure 4.3 Base fixity

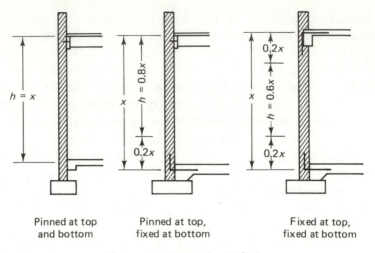

| Pinned at top
and bottom | Pinned at top,
fixed at bottom | Fixed at top,
fixed at bottom |

Figure 4.4 Vertical wall fixity

point and the analytical model assumed by the designer is no longer valid.

If a wall is assumed to span from pilaster to pilaster or wall to wall, the effective height is found in the same manner as that for a multistory wall. The points of inflection are assumed at $0.2x$ and the value of h is then equal to $0.6x$. The restraint provided by end walls is generally sufficient, so that the point of inflection should still fall somewhere around $0.2x$ from the end wall. Figure 4.5 shows the

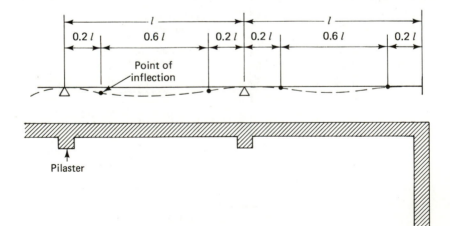

Figure 4.5 Horizontally spanning wall fixity

assumed points of inflection in a wall spanning horizontally between vertical supports.

When pilasters are loaded by horizontally spanning walls the total amount of load that is assumed to load the pilaster is smaller than that calculated using the model shown in Figure 4.5. This situation arises because pilasters are not infinitely rigid as the model assumes and the wall tends to behave like a two-way slab.

The National Concrete Masonry Association has published [4.1] a set of curves that try to account for the distribution of transverse lateral loads to the pilasters in walls subject to two-way action. The curves shown in Figure 4.6 estimate the percentage, k, of lateral load carried by the pilaster for various wall fixity conditions. The revised lateral load w' that the pilaster must carry is

$$w' = kwl \qquad (4.1)$$

where k = coefficient from Figure 4.6

w = design transverse lateral load applied to panel

l = center-to-center span between pilasters

4.4 STABILITY CONSIDERATIONS

Maximum span-to-thickness (h/t) ratios are specified for all material used to support axial load. Masonry is not excepted; Table 4.1 is taken from the UBC. The sense of height refers to the governing span, either between floors or pilasters. This h/t ratio varies with masonry type, loading condition, and whether or not the wall behaves like a deep beam. In addition, the code specifies the minimum nominal block thickness of the wall. This h/t limitation superficially appears to originate from stability considerations, when in fact it is a serviceability requirement. A wall may have sufficient strength to resist the design loads, but may not be considered serviceable in that it may, for example, experience undesirable cracking at typical service loads. Serviceability is an important consideration in the design of any structure and certainly an important consideration in the design of walls because one of their primary functions is to keep wind and water out of a building. Controlling serviceability with h/t ratios, while appropriate for the design of houses and small industrial and small commerical buildings, is not appropriate for the design of large structures. In this case cracking moments should be

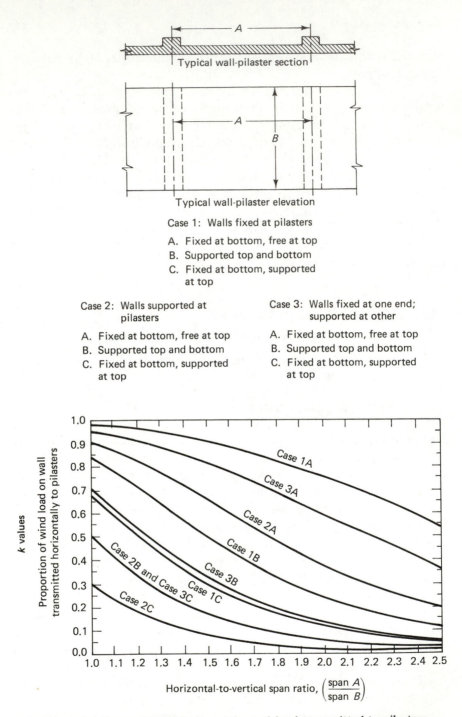

Typical wall-pilaster section

Typical wall-pilaster elevation

Case 1: Walls fixed at pilasters

A. Fixed at bottom, free at top
B. Supported top and bottom
C. Fixed at bottom, supported at top

Case 2: Walls supported at pilasters

A. Fixed at bottom, free at top
B. Supported top and bottom
C. Fixed at bottom, supported at top

Case 3: Walls fixed at one end; supported at other

A. Fixed at bottom, free at top
B. Supported top and bottom
C. Fixed at bottom, supported at top

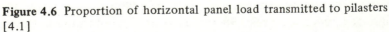

Figure 4.6 Proportion of horizontal panel load transmitted to pilasters [4.1]

TABLE 4.1   Unsupported Height-to-Thickness Ratio h/t [4.2]

Type of Masonry	Maximum h/t	Minimum Nominal Thickness (in.)
Reinforced grouted masonry (bearing)	25	6
Reinforced grouted masonry (bearing) and designed as a deep beam	36	6
Exterior reinforced walls (nonbearing)	30	2
Interior reinforced walls (nonbearing)	48	2

used as a guide to serviceability as they are for reinforced concrete members. How, then, are stability concerns to be addressed?

Walls loaded in pure axial compression will fail when compressive crushing of the masonry occurs unless the axial capacity of a member is limited by stability considerations. Consequently, an axially loaded wall may fail from buckling at a compressive stress level significantly below that predicted by strength considerations. The tendency to fail because of lateral instability is greater in slender walls than it is in stouter members, such as columns, primarily because unintentional deviations from plumb have more impact. Euler's equation of elastic buckling gives the stress at which a tall, thin member will fail:

$$f_{m \, \max} = \frac{\pi^2 E_m}{(kh/r)^2} \qquad (4.2)$$

where E_m = elastic modulus of concrete masonry

   h = unbraced length

   r = least radius of gyration

   k = effective height coefficient

Note that this critical stress is not a direct function of material strength.

In Equation (4.2) the coefficient k establishes the effective height of the wall by accounting for end restraint conditions. The values of k, shown in Table 4.2, numerically modify the measured unbraced length so as to produce an equivalent simply supported height. It should be noted that there are two values of k in Table 4.2 for each fixity condition. The theoretical value is obtained by solving the governing differential equations for the assumed support conditions and is nonconservative. The recommended value of k

TABLE 4.2 k Factors [4.3]

	(a)	(b)	(c)	(d)	(e)	(f)
Buckled shape of column is shown by dashed line						
Theoretical k value	0.5	0.7	1.0	1.0	2.0	2.0
Recommended design value when ideal conditions are approximated	0.65	0.8	1.2	1.0	2.1	2.0
End condition code		Rotation fixed and translation fixed				
		Rotation free and translation fixed				
		Rotation fixed and translation free				
		Rotation free and translation free				

attempts to account for various construction inaccuracies and a variation between the analytical model and actual construction conditions such as the degree of rotational restraint actually provided. The values shown in Table 4.2 are taken from the American Institute of Steel Construction Specifications [4.3] for the design of steel structures; a set of k values for masonry structures should be developed, but at the present time professional experience must be utilized.

The expression developed by Euler tends to overestimate column capacity because it presupposes an elastic material with a uniform modulus of elasticity, and this is not the case for concrete masonry. Inelastic buckling occurs in steel members and results in lateral instability at loads lower than those predicted by Euler (Figure 4.7). More data are needed to define the transitional equation to account for this type of behavior in masonry; however, the

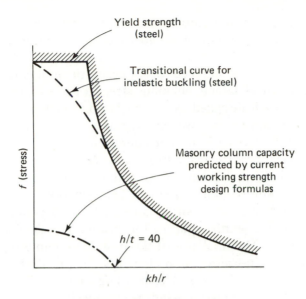

Figure 4.7 Buckling curve

approach adopted in explaining the same phenomenon in steel will undoubtedly have its parallel in concrete masonry.

The curve shown in Figure 4.7, often referred to as the *Johnson–Euler curve*, shows that the critical stress for a steel member is related to elastic and inelastic buckling, and the ultimate strength of the member. A representative curve showing the allowable stresses for masonry using the present code approach is also presented in the figure. Note that the shapes of the steel and masonry curves in the elastic region are significantly different. The steel curve is concave upward whereas that for masonry is concave downward, contrary to the direction suggested by Euler. There currently is no reliable expression to accurately quantify the strength of masonry in axial compression when stability is a consideration. Research in concrete members subjected to axial loads suggests that it is important to utilize the appropriate model for establishing member cross-sectional properties when evaluating stability. The use of gross cross-sectional properties is generally an inappropriate model because of the possibility of large member deflections with the accompanying reduction of member stiffness. However, recent tests by the Structural Engineers Association of Southern California [SEAOSC] [4.4] conducted in Los Angeles indicate that stability is not as important a consideration as once thought.

Walls that are not supported on a continuous footing, but instead span between individual pad footings placed at intervals, are modeled as deep-wall beams, as shown in Figure 4.8. It has been shown that vertically induced deep beam flexural stress patterns increase the critical lateral buckling stress for a wall and hence the h/t ratio for a wall designed as a deep beam is increased from 25 to 36. This is, of course, inconsistent with the conceptual development of h/t ratios (i.e., serviceability considerations).

A deep-wall beam spans between the two footings with the total length of wall supported on the footing being no greater than one-tenth of the center-to-center spacing between the footings (Figure 4.8). The reaction at the footing is carried by the wall as an in-wall column and has a width equal to the bearing length plus twice the block thickness. The thickness of the in-wall column is equal to that of the wall. The in-wall column elements must be designed to resist the vertical axial reaction as well as bending stresses.

Most masonry walls modeled as deep-wall beams do not span between the footings as simply supported beams but rather as continuous elements. Loads are usually applied to the top edge of the wall in the form of uniformly distributed loads and point loads. It is clear that the wall must be reinforced to resist the resultant tensile stresses at the midspan as well as those over the supports.

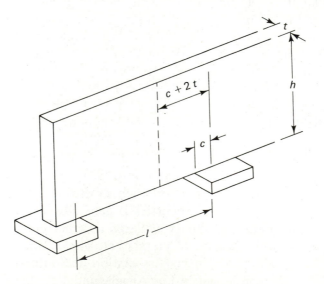

Figure 4.8 Deep-wall beam

4.5 MODELING THE DIAPHRAGM

Floor and roof diaphragms are a very important part of the overall lateral bracing scheme for a building. They transfer inertia loads from walls and floor systems to the vertical bracing elements of the building. As a result they are, in effect, horizontal beams. Most often the shape of a diaphragm is square or very nearly square and the appropriate beam model is that of a beam in which shear deformation adequately characterizes the behavior of the system. Typically, the beam model consists of a web which takes only shear loads, and collector chords and struts which take axial components of the load. The web or diaphragm is usually concrete, concrete and metal deck, metal deck, or plywood. Chord and collector forces are taken in steel beams, bond beams in walls, or reinforcing in concrete topping. Diaphragm flexibility and load path are the main engineering considerations in attempting to model the diaphragm. They will be discussed separately.

Engineers have traditionally characterized diaphragms by the type of material from which they are constructed. A plywood diaphragm is a "flexible" diaphragm, a concrete diaphragm is a "rigid" diaphragm, and so on. This model is developed based on an understanding of how the system behaves in its usual application.

A flexible diaphragm distributes inertia loads to supporting shear walls based on the tributary width "supported" by each shear wall in the same way as gravity loads are allocated to supports from pin-connected or simply supported beams. Figure 4.9 shows the tributary load which is carried by each shear wall in a box system with a flexible diaphragm. The rationale for assigning loads to the shear walls in this manner is that the diaphragm is considered to be much more flexible than the shear walls and no appreciable continuity is assumed to exist over the supports. The diaphragm is then modeled as several simply supported beams, with the load between two adjacent shear walls split evenly between them.

Although shear walls II and III in Figure 4.9 are loaded by equivalent tributary widths, the shear stress that wall II must resist is greater than that resisted by wall III. The shorter length of wall II requires that a smaller shear area resist the same shear load as that applied to wall III.

The rationale for this modeling procedure is well based in theory. The deflection of the diaphragm is much greater than the

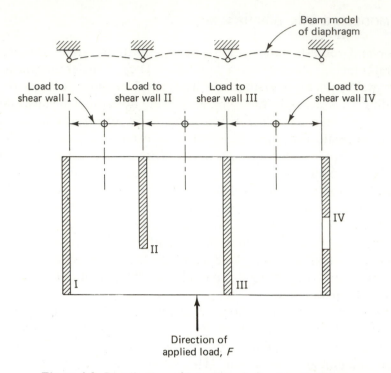

Figure 4.9 Distribution of lateral load: flexible diaphragm

wall deflection and the support deformation can be assumed to be negligible. Continuity effects may be disregarded, as shear deformation (a sliding of planes) usually accounts for the majority of the diaphragm's flexibility. The simply supported beam model adequately represents system behavior and can be confidently used as the analytical model.

Sometimes the designer forgets the theoretical basis for the everyday models and tends to force designs that are inconsistent with the way the structure will behave. For example, as wall II in Figure 4.9 is shortened, its stiffness is reduced and the flexible diaphragm concept must be used with caution. *The stiffness of the shear wall and diaphragm must be compared to determine the appropriate load distribution.* Elastic analysis does not shed much real insight into the problem because predicting real deflections depends, to a large extent, on the load level since both systems are nonlinear, especially when subject to loads generated by design-level earthquakes. The conservative approach requires the designer to discount almost entirely the contribution of a bracing wall that is considerably shorter than the other bracing components of the system. The justification lies in the rotation that will probably occur in the foundation or in wall ductility. Figure 4.10 replaces wall II with a spring whose force/deformation characteristics can be approximated based on the characteristics of the wall. The diaphragm can now be designed for this yield level load, with the remaining loads going to walls I and III. A conservative estimate of force is appropriate for estimating diaphragm forces; however, load distributing elements such as drag struts or collectors should be proportioned to account for nonconservative wall strengths and stiffness.

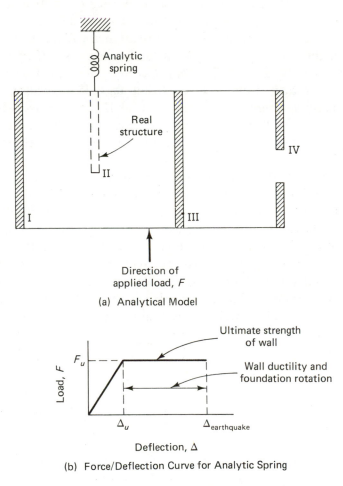

(a) Analytical Model

(b) Force/Deflection Curve for Analytic Spring

Figure 4.10 Analytical model

Concrete diaphragms and metal deck diaphragms are generally classified as "rigid" or in the case of metal deck sometimes by the confusing term "semirigid." These diaphragms are presumed not to deform in plane when subjected to anticipated earthquake loads, or more specifically that their deformation will be significantly less than that of the associated vertical bracing system. As a result, load will be distributed to vertical bracing in direct proportion to the stiffness of the vertical bracing elements. Once again this is the usual model and is appropriate for most designs; however, the designer must be aware of the conditions under which this model becomes inappropriate.

Consider, for example, the diaphragm depicted in Figure 4.11. Torsional stiffness is provided by walls III and IV. How effective these walls are in providing torsional resistance will have a significant effect on the load that wall II must be able to resist. If they provide full torsional restraint, wall II will take very little load because it is more flexible than wall I. On the other hand, if the walls III and IV or the diaphragm are quite flexible, wall II must carry half of the building shear. The engineer must now carefully consider diaphragm flexibility as well as foundation rotation to model the system properly.

Buildings with rigid diaphragms respond to dynamic input with both translational and torsional modes of vibration. Torsional modes of vibration are not usually a critical element in the design of low-rise buildings; however, torque generated during translational modes of vibration is a major concern. Figure 4.12 illustrates the situation when the force acts through the center of mass, but the building rotates about the center of rigidity. This creates torsion in the structure, which produces higher force levels in some walls and a reduction in forces in other walls. The reduction in force associated with

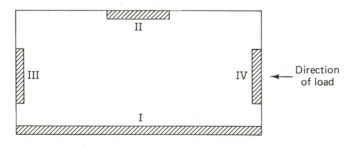

Figure 4.11 Distribution of lateral load: rigid diaphragm

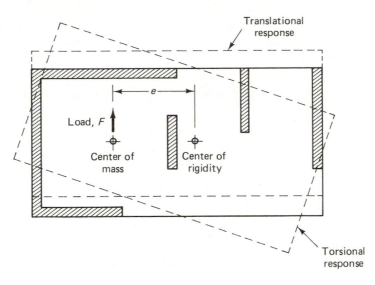

Figure 4.12 Torsional and translational response

torsion is neglected because of the cyclic nature of earthquake load-ings (i.e., load reversals) and the consequences of an inappropriate selection of the analytical model. Notice that this conservatism does not mitigate the selection of an inappropriate model in the example discussed previously.

Torsion is present in most structures, even those where the center of mass and center of rigidity theoretically coincide. This accidental torsion may occur because of modeling assumptions, construction inaccuracies, or because torsional modes of vibration may be close to the dominant lateral modes. The building code requires that the designer make provision for obvious geometric configurations which create torsion, and it also specifies a minimum torsional eccentricity (e) of 0.05 times the maximum planar build-ing dimension at the level under consideration to account for this accidental torsion.

The center of mass is established in the same manner as the centroid of a structural section. The center of rigidity is located by calculating the rigidity of each wall and imposing a linear deforma-tion pattern on the presumed rigid diaphragm. The magnitude of the torsional moment is determined by applying the shear force at an eccentricity equal to the distance between the center of mass and the center of rigidity.

Diaphragms, both rigid and flexible, are modeled as though the web of the diaphragm is not capable of taking bending stress and that the chord elements, located at the perimeter, collect the imposed shear forces. Loading, shear flow, and shear forces are as depicted in Figure 4.13. One may calculate the chord force, for example at T_A,

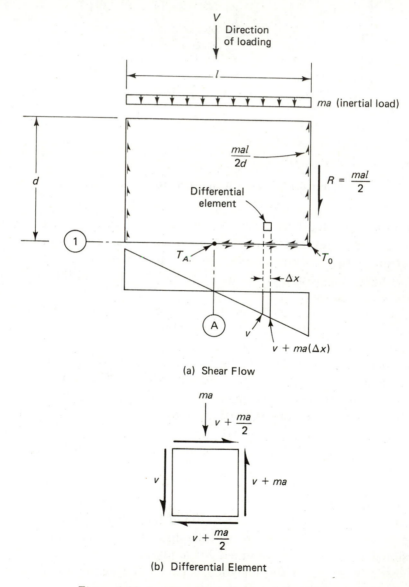

(a) Shear Flow

(b) Differential Element

Figure 4.13 Loading, shear flow, and shear force

by combining unit shears imposed on the collector chord by the diaphragm. Hence.

$$T_A = \text{(average shear flow on line 1 between } T_A \text{ and } T_O\text{)} \times \frac{l}{2}$$

$$= \left(\frac{ma\,l}{2d}\right)\left(\frac{1}{2}\right)\left(\frac{l}{2}\right) = \frac{(ma)\,l^2}{8d} \tag{4.3}$$

where ma is the percentage of tributary earthquake load (similar to the $C_p W$ concept used in the *Uniform Building Code*; refer to Volume 1). This is the same conclusion as that which can be developed by dividing the moment at A by the distance between chords:

$$T_A = \frac{M_A}{d} = \left(\frac{ma\,l^2}{8}\right)\left(\frac{1}{d}\right) = \frac{(ma)\,l^2}{8d} \tag{4.4}$$

Since the latter approach is more familiar to the engineer, it is used most often; however, the approach used in Equation (4.3) is more descriptive of the shear flow in a diaphragm and more readily explains how chords and struts are loaded.

To illustrate, assume that the following is a design consideration. Line 1 is to be built as shown in Figure 4.14. The usual problem is to define the force criteria for axial loads imposed on the beam by earthquake forces. If we use the model discussed previously for the diaphragm consisting of a web and a chord, the flange force

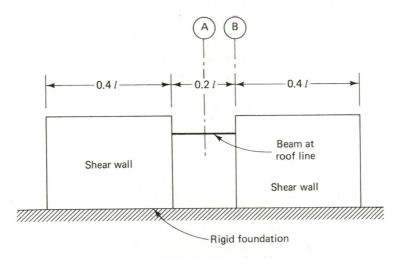

Figure 4.14 Building elevation

can be calculated by either Equation (4.3) or (4.4). Equation (4.4) becomes inappropriate, however, with the introduction of the stiff shear walls shown in Figure 4.14, where the maximum chord force imposed on the beam element is only that for shear flow which occurs between points A and B of Figure 4.14.

$$T_A = \text{(average shear flow between } A \text{ and } B) \times l_{AB}$$

$$= (0.1l)\left(\frac{ma}{d}\right)\left(\frac{1}{2}\right) \ (0.1l) = (0.005)(ma)\left(\frac{l^2}{d}\right) \qquad (4.5)$$

This is only 4 percent of the force suggested by Equation (4.4). Two points are important:

1. Clearly, the stiff shear wall transfers the chord forces directly to the ground, thereby making a chord at the roof line for this type of system redundant.

2. The force development of Equation (4.3) makes it easier to derive forces acting on components. The governing criteria for the beam of Figure 4.14 will be the collector forces generated by seismic loading in the orthogonal direction.

$$T_{\max} = \left(\frac{mal}{2d}\right)(0.1l) = \frac{0.05(ma)\,l^2}{d} \qquad (4.6)$$

This force will be a maximum at the face of the wall [this analysis assumes that the seismically induced load (ma) is the same in each direction].

Understanding shear flow in a diaphragm as described in the development of Equation (4.3) is essential to the development of load criteria for the design of collectors.

Figure 4.15a shows the relative deflections of a flexible diaphragm in a building with an irregular shape if the diaphragm were not tied together. For the purposes of illustration, the diaphragm is divided into three areas as shown. It can be seen that the diaphragm in area II is deflecting differently from the diaphragms connected to it. The diaphragms in areas I and III, in fact, have no significant deflections at the interior points where they are connected to the diaphragm in area II because the shear walls to which they are connected are much stiffer than the big diaphragm. Behavior such as this is undesirable during an earthquake, and a method of assuring compatible deflections between the different diaphragm areas must be employed. This is typically accomplished by tieing the different diaphragms into a single unit through the use of collector elements.

These collectors act in tension or compression (depending on the direction of the lateral loading) to tie the diaphragm together at points of discontinuity or change in plan direction, and transfer load into the shear wall to which they are connected. The big diaphragm, now divided into three regular diaphragms (Figure 4.15b), can now act as a single unit without the disruptive tearing forces inhibiting the successful performance of the system. The collector must be designed for the appropriate axial load, and the connection of the collector to the wall must be sized to develop the capacity of the collector.

For the flexible diaphragm shown in Figure 4.15a, the shear distribution along the collector is shown in Figure 4.15b. The maximum collector force is the sum of the shear flows loading the collectors.

$$T_x = \left(\frac{m_1 a l_1}{2d_1}\right)(d_1) + \left(\frac{m_2 a l_2}{2d_2}\right)(d_1) \qquad (4.7)$$

Shear distribution is different for a rigid diaphragm, Figure 4.15c, as the wall on line A will carry less load than the wall on line B. This is contrasted with the load distribution which occurs based on tributary areas when the diaphragm is considered flexible. The diaphragm between A and B will carry a portion of the shear stress generated by the diaphragm of span l_1 which is now redistributed by the collector at line A. This is the case when R_1 is less than $m_1 a l_1 / 2$. The shear distribution described is possible only if a shear resisting element is available to take diaphragm loads along line 1 between A and B. Should a shear collector not be available at line 1, shear distribution to the lower part of the diaphragm would not be possible since perimeter shear forces at line 1 could not be resisted. The diaphragm between lines 2 and 3 would now resist all lateral loads and a collector or chord would now be required along line 2. The shear flow in the diaphragm at line B between lines 2 and 3 will be

$$Q_{2\text{-}3} = \frac{m_1 a l_1}{2d_1} - \frac{R_1}{d_1} + \frac{m_2 a l_2}{d_1} \qquad (4.8)$$

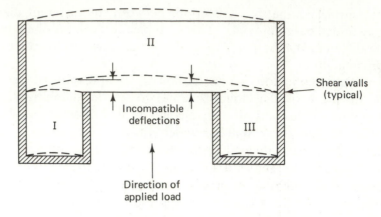

(a) Irregular Flexible Diaphragm with Incompatible Deflections

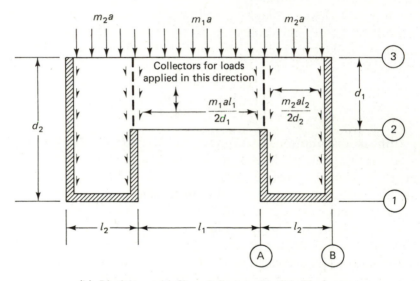

(b) Diaphragm with Shear Collectors: Flexible Diaphragm

Figure 4.15 Diaphragms with irregular plans

Treatment of flexible disphgram shear flow around an opening is quite similar to that developed in Figure 4.15.

The first task is to select appropriate values for the mass distribution (m_a, m_b, and m_c) shown in Figure 4.16. Once this has been done, shear flows can be developed. Elemental shear flows are shown in Figure 4.16 along line 3. Collector forces are calculated based on this assumed distribution of shear flow:

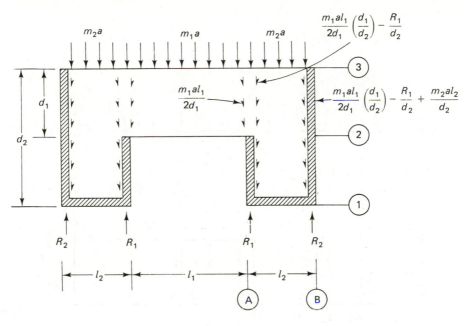

(c) Rigid Diaphragm

Figure 4.15 (*Continued*)

At w:

$$T_w = \text{(average shear flow) (length)}$$

$$= \left[\frac{m_3 a l_1}{2d_3}\left(\frac{1}{2}\right)\right]\frac{l_1}{2} \tag{4.9}$$

$$= \frac{m_3 a l_1^2}{8d_3}$$

At x along line C:

$$T_x = 0$$

At x along line 3:

$$T_x = \frac{m_3 a l_1}{2d_3}(d_3) - d_3\left[\frac{m_1 a l_1}{2d_1}\left(\frac{d_1}{d_1 + d_2 + d_3}\right)\right.$$

$$\left. + \frac{m_3 a l_1}{2d_3}\left(\frac{d_3}{d_1 + d_2 + d_3}\right)\right] \tag{4.10}$$

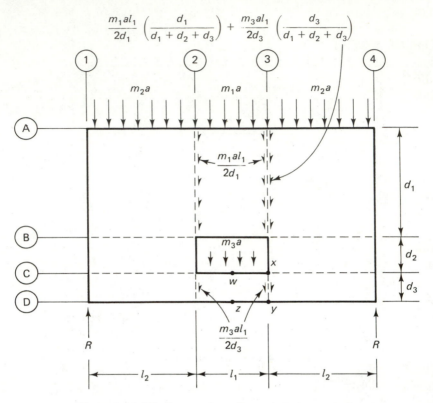

Figure 4.16 Diaphragm shear flow around an opening

At y:

$$T_y = \left[\frac{m_1 a l_1}{2d_1} \left(\frac{d_1}{d_1 + d_2 + d_3} \right) + \frac{m_3 a l_1}{2d_3} \left(\frac{d_3}{d_1 + d_2 + d_3} \right) \right.$$
$$\left. + \frac{m_2 a l_2}{2(d_1 + d_2 + d_3)} \right] \tag{4.11}$$

At z:

$$T_z = T_y + \left[\frac{m_3 a l_1}{2d_3} \left(\frac{1}{2} \right) \frac{l_1}{2} \right] \tag{4.12}$$

Although these equations appear complex, the derivations are quite straightforward, as can be seen in the preceding development.

4.6 MODELING THE WALL FOR IN-PLANE LOADS

Once it has been determined what amount of load a shear wall or system of shear walls is expected to be capable of withstanding, the allocation of this load must be made. Two problems are typical of this topic area:

1. Force distribution by a rigid diaphragm

2. Force distributions to components of a wall with openings or discontinuities

Both problems require the designer to determine appropriately the relative rigidity of the masonry walls.

The *rigidity* of a wall is defined as the reciprocal of its deflection when subjected to a unit load. Rigidity is a function of boundary conditions, physical dimensions, and the characteristics of the material. Insofar as determining the relative rigidity of a system of shear walls is concerned, it is not the actual rigidity or stiffness that is of real interest but the relative rigidity of each wall with respect to the other walls. Walls that are fixed at the top and bottom will deflect a different amount than an otherwise identical wall that is fixed at the bottom and pinned at the top. Walls with different heights, thicknesses, and widths will also behave differently. The modulus of elasticity, E_m, and the shear modulus, E_v, must be established in order to calculate the deflection of the walls, although for walls built using the same material, these factors may be ignored in determining the relative rigidity of the shear walls.

Figure 4.17 shows the two common models used to establish the stiffness of a shear wall. A wall assumed to be fixed at the top and bottom is modeled as if it were a fixed-fixed beam, while a wall fixed at the bottom and pinned at the top is modeled as a cantilevered beam.

The deflection of a shear wall is the sum of the deflection caused by bending deformation, Δ_b, and those caused by shear deformation, Δ_v. The general equation for the deflection of a wall modeled as a fixed-fixed beam is

$$\Delta_{F\text{-}F} = \Delta_b + \Delta_v \qquad \text{(fixed-fixed)}$$

$$= \frac{Vh^3}{12E_m I} + \frac{1.2Vh}{AE_v}$$

(4.13a)

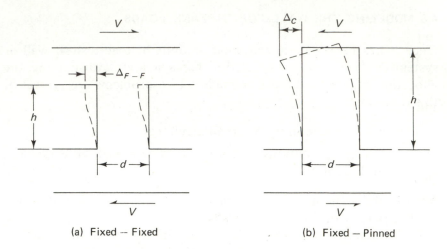

(a) Fixed -- Fixed   (b) Fixed -- Pinned

Figure 4.17 Shear wall deflection

and that for a wall modeled as a cantilever beam is

$$\Delta_C = \Delta_b + \Delta_v \qquad \text{(cantilever)}$$

$$= \frac{Vh^3}{3E_m I} + \frac{1.2Vh}{AE_v} \qquad (4.13b)$$

where V = applied shear

I = gross moment of inertia

A = gross cross-sectional area in the horizontal plane

E_m = modulus of elasticity

E_v = shear modulus

If E_m and E_v are constant for all the walls, and E_v is taken as $0.4E_m$, Equation (4.13) may be modified to express the relative deflection per unit of thickness in terms of the more convenient h/d ratio (as shown in Figure 4.17).

$$\Delta_{F\text{-}F} = \frac{V}{E_m} \left[\left(\frac{h}{d}\right)^3 + 3\left(\frac{h}{d}\right) \right] \qquad \text{(fixed-fixed)} \quad (4.14a)$$

and

$$\Delta_C = \frac{V}{E_m} \left[4\left(\frac{h}{d}\right)^3 + 3\left(\frac{h}{d}\right) \right] \qquad \text{(cantilever)} \quad (4.14b)$$

The rigidity of the wall model above is the force, R, required to produce a unit of displacement:

$$R_{F\text{-}F} = \frac{1}{\Delta_{F\text{-}F}} \qquad \text{(fixed-fixed)} \qquad (4.15a)$$

and

$$R_C = \frac{1}{\Delta_C} \qquad \text{(cantilever)} \qquad (4.15b)$$

Quite often a shear wall actually consists of a combination of several walls in the same plane or a single wall with openings, and the model must account for the fact that the shear wall consists of these several components.

In examining the components of Equations (4.14a) and (4.14b) it can be seen that they each consist of two terms containing (h/d) and $(h/d)^3$. The cubic part of the equations represents bending deformation and the linear part shear deformation. For (h/d) values in excess of 1.0 bending tends to dominate, while for h/d values less than 1.0 shear will tend to dominate. Figure 4.18 shows a wall that might be a part of the bracing system for a one-story building. In

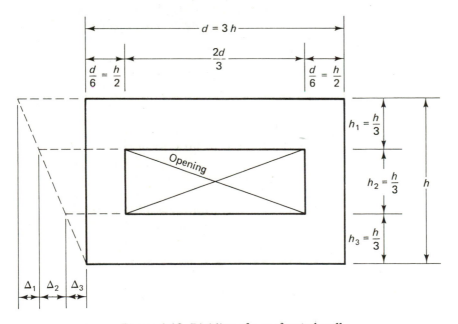

Figure 4.18 Rigidity of a perforated wall

this case, the wall will behave like a cantilever beam and use of Equation (4.14b) to calculate the relative deflection is appropriate.

If the opening is ignored and it is assumed that V/E_m equals unity, then the relative contribution of bending and shear deformation to the total wall deflection can be examined. In a one-story building such as that shown in Figure 4.18, the h/d ratio is often less than 1.0. If it is assumed that $d = 3h$ (i.e., $h/d = 0.33$), the relative lateral deflection of the wall is

$$\Delta_C = 4(0.33)^3 + 3(0.33)$$
$$= 0.144 + 0.99 \tag{4.16}$$

Clearly, little accuracy would be lost if bending stiffness were neglected. The rigidity of the wall, R_W, is

$$R_W = \frac{1}{\Delta_C} = \frac{1}{0.99} = 1.01 \tag{4.17}$$

(R_W is 0.905 if the bending term is included.)

If the opening shown in Figure 4.18 is now considered, determining the stiffness of the wall becomes more complicated. Various methods are commonly used to determine an appropriate *estimate* of wall stiffness; however, building on our previous work it seems appropriate to evaluate the stiffness by measuring the shear deformation above and below the opening and the deformation of the piers at the level of the opening.

$$\Delta_1 = 3\left(\frac{h/3}{3h}\right) = 0.33 \tag{4.18}$$

$$\Delta_3 = \Delta_1 \tag{4.19}$$

$$\Delta_2 = \Delta_{F\text{-}F} = \left[\left(\frac{h/3}{h/2}\right)^3 + 3\,\frac{h/3}{h/2}\right]\left(\frac{1}{2}\right)$$
$$= 1.15 \tag{4.20}$$

$$\Delta_W = 0.333 + 1.15 + 0.333 = 1.81 \tag{4.21}$$

$$R_W = \frac{1}{\Delta_W} = \frac{1}{1.81} = 0.551 \tag{4.22}$$

In equation (4.20), $V = \frac{1}{2}$ because there are two piers.

In this case, the stiffness of the wall is to a large extent a function of the stiffness of the piers on either side of the opening. Designers

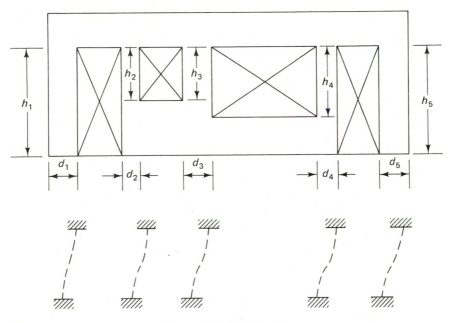

Figure 4.19 Wall rigidity

will commonly neglect the deformation of the wall segments above and below the windows, and an appropriate model for the wall shown in Figure 4.19 becomes one of a series of fixed-fixed piers.

The rigidity of the entire wall, R_W, must be determined by imposing a unit deformation on each of the piers and estimating the force required to accomplish this deformation. The rigidity of the wall is the sum of the individual wall rigidities:

$$R_W = R_1 + R_2 + R_3 + R_4 + R_5$$

$$= \frac{1}{(h_1/d_1)^3 + 3(h_1/d_1)} + \frac{1}{(h_2/d_2)^3 + 3(h_2/d_2)} + \cdots$$

(4.23)

Once the rigidities of all of the individual walls or piers have been established, the rigidity of each wall with respect to the other walls in the system can be established. This relative rigidity, $\overline{R}$, is given by

$$\overline{R}_i = \frac{R_i}{\sum\limits_{i=1}^{n} R_i}$$

(4.24)

where R_i denotes the rigidity of the ith shear wall.

This modeling concept reduces a complex problem to one that is easily handled by the designer; however, it must be understood when the assumptions used to develop the model are no longer valid. For example, if the opening in Figure 4.18 is reduced from $0.67d$ to $0.2d$ and the modeling technique shown in Figure 4.19 is used, the apparent wall stiffness becomes

$$R_W = 2\left[\dfrac{1}{\left(\dfrac{h/3}{1.2h}\right)^3 + 3\left(\dfrac{h/3}{1.2h}\right)}\right]$$

$$= 2.34 \tag{4.25}$$

This model wall with the opening is stiffer than the equivalent wall without an opening [this implies that the previous value of R_W from Equation (4.15) was 1.01]. Of course, this is not possible because the model for the wall was inappropriately chosen using Equation (4.25).

Poor or inappropriate modeling of building components can lead to erroneous conclusions or require the designer to expend a considerable amount of extra analytical effort. A good basic understanding of deformation analysis and system behavior is essential to the development of component models.

4.7 COMPONENT MODELING: GENERAL

Once the force to which a component will be subjected has been estimated, the effect of this force on the component must be evaluated. The development of component models is the result of extensive behavioral test programs. This is especially true for structural steel and reinforced concrete. Other building materials rely on a more limited data base to develop component models; concrete masonry is such a building material. Forty years ago reinforced concrete relied on a elastic model to evaluate the behavior of its components. Today, after an extensive series of test programs, concrete components are analyzed using a variety of models. The appropriate analytical model depends on the load level to which the member is subjected. At low force levels, the material behaves as though it were a homogeneous material and the "uncracked" section is used to model the beam. Once the force level passes above the range in which tension in the concrete can be assumed to carry a portion of the load, a cracked section becomes the model. This cracked section discounts the material in the tension region and derives its strength from the reinforcing steel and the compression portion of the member. As the load and deformation increase, the model for the compression zone is, once again, changed as the stress-strain relationship for the concrete changes. The appropriate model for the compression region then becomes the Whitney stress block.

Concrete masonry today still uses the elastic model to evaluate the behavior of its components, even though engineers generally agree that its behavior should be similar to that of concrete. The depth of the stress block and stress distribution significantly affect structural behavior, and concrete masonry will undoubtedly have some modeling considerations which are different from those used in concrete design because one is now dealing with a more complex composite material. The effect of block size, lay-up, and grouting may have a significant impact on the analytical model. More test data must be obtained, but it should not preclude the use of engineering judgment in the development of an appropriate model. Tests which thoroughly study the impact of variables on modeling assumptions, especially when subjected to high levels of stress and strain, must be made. The sections that follow deal with component modeling. Appropriate component models for members in flexure and compression will be developed and stability considerations addressed.

4.8 COMPONENT MODELING: MEMBERS SUBJECTED TO FLEXURAL LOADS

The mechanics of a member subjected to flexural loads have been discussed in Chapter 3. Now the appropriateness of this analytical treatment to concrete masonry must be established. An analytical model must be developed that links this theory with observed behavior. Since the relationship between stress and strain in concrete and concrete masonry is a function of load level, various models are required to describe properly the behavior of a concrete masonry member. At low stress levels concrete and concrete masonry can carry tensile loads; hence, the appropriate homogeneous model must be developed. As the load level increases, the tensile strength of the material is soon exceeded and reinforcing steel is relied on to carry the tensile forces. Now a cracked section is required. As compressive stresses become large, the relationship between stress and strain becomes nonlinear and the model again changes.

Wall construction constitutes the largest class of concrete masonry construction, and therefore this section focuses on wall design considerations. If a wall section is to serve its purpose, the potential for flexural cracks should be minimized at load levels that are expected to occur repeatedly during the life of a structure. This is commonly referred to as a *serviceability criterion*. Cracks will begin to form in mortar joints when the tension stress in the mortar exceeds the cracking stress. An analytical model of the wall must be developed in order to estimate the tensile strength of the section. The model developed must account for the type of construction proposed. Since test data are limited, engineering intuition must serve as a guide to proper model development. From a serviceability standpoint, two types of wall construction must be considered: partially grouted walls and fully grouted walls.

In a fully grouted wall, the uncracked section properties can be estimated by assuming that the walls behave like solid slabs because the mortar and grout interlock at each bed joint to make composite action possible. The effective thickness of the wall, h_{eff}, should consider the effect of joint treatments, and if the bed joints are deeply tooled or raked, deducting $\frac{1}{4}$ in. from each face thickness is probably realistic. Figure 4.20a shows how joint treatments affect the effective thickness of the wall. For a partially grouted section, the mortar joint becomes the chord, and the appropriate model is shown in Figure 4.20b. In the partially grouted uncracked section the joint thickness, t, is now the block shell thickness less any joint treatment (depth of rake). Since the flanges (mortar joints) are connected by the block webs the spacing of the vertical reinforcing (grouted cell) or the lay-up of the block should have no effect on the effective flange width. Ignoring the moment of inertia of the joint about its own axis and that of the web (because there is often no mortar at the web to transfer tension or compression), the effective uncracked section modulus, S, for the partially grouted wall is

$$S = \frac{2\left[bt\left(\dfrac{h_{\text{eff}}}{2} - \dfrac{t}{2}\right)^2\right]}{h_{\text{eff}}/2}$$

$$= \frac{bt}{h_{\text{eff}}}(h_{\text{eff}} - t)^2 \tag{4.26}$$

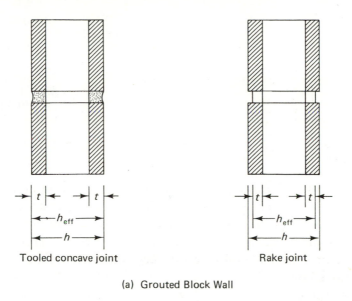

Tooled concave joint   Rake joint

(a) Grouted Block Wall

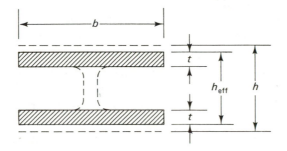

(b) Ungrouted Block Wall

Figure 4.20 Section properties—uncracked section

Tests have shown that the tensile strength of concrete masonry construction has a lower bound of approximately 100 psi [4.5]. Tensile stresses in the mortar of 100 psi or less should produce no cracking in the wall.

Reinforcing steel in masonry walls located near the center of the cell is significantly strained only after the mortar or concrete has cracked. The appropriate component model for the wall is now derived from the compressive and tensile components of the uncracked section (i.e., neglecting those portions that are subjected to tensile strains). Once again the presence or absence of grout affects the shape of the compressive stress block, and the models for a concrete

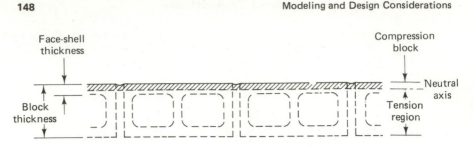

Figure 4.21 Compression block totally within face shell (rectangular beam)

masonry wall resemble that used for a rectangular beam or a T beam. In thin-wall sections, the depth of the compressive stress block is typically less than the thickness of the face shell, and a rectangular section is appropriate (see Figure 4.21). Partially grouted walls thicker than 8 in. and doubly reinforced walls may have compressive stress blocks that extend into the grouted cell (see Figure 4.22). In both cases the effective flange width, b, becomes a consideration. The UBC limits the value of b to the smaller of the center-to-center spacing of the vertical steel or six times the block thickness for running bond or three times for stack bonded walls. The more restrictive requirement for stacked bond results from the assumption that the continuous vertical mortar joint in stack bond walls (head joint) is less effective in transferring the stresses laterally along the wall than the staggered vertical joint present in running bond construction.

The Masonry Institute of America [4.6] conducted a series of tests to determine the effective width, b, of a wall in flexure. In examining walls made with 6- and 8-in. block in a running bond, it appeared that an 8-ft spacing of the vertical steel was as effective in

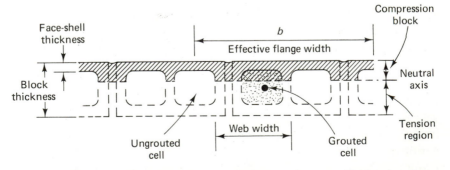

Figure 4.22 Compression block falls in grout space (T beam)

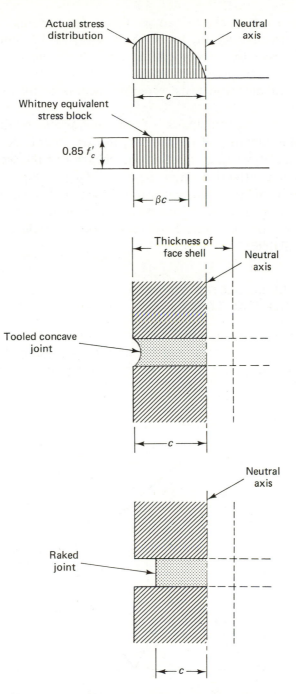

Figure 4.23 Whitney stress block at mortar joint

resisting the out-of-plane load as was a 2-ft spacing. This would imply an effective width, for an 8-in. wall, of 12 times the block thickness—twice that specified in the UBC. The 8-in. block wall built in a stacked bond was about half as effective as the running bond. Based on these tests, effective widths of up to 48 in. seem reasonable.

The effective thickness of the compression flange should consider the thickness of the mortar joint treatment. Normal tooling probably has little or no effect on the behavior of the section in compression since effective stress near the face is discounted in the development of the Whitney stress block and a slightly concave mortar joint probably performs much like a flush joint (Figure 4.23). A deep rake, on the other hand, will probably reduce the effective section available.

4.9 MEMBERS SUBJECTED TO AXIAL AND FLEXURAL LOADS

Theoretical treatments used to predict the behavior of concrete sections subjected to axial and flexural loads are complex. Practical design procedures were developed by Whitney and are based on ob-

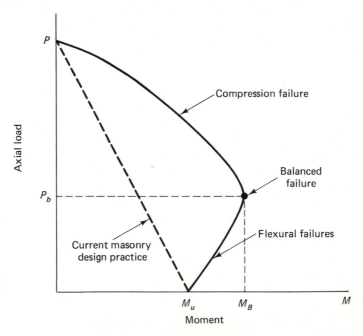

Figure 4.24 Interaction curve

served behavior of test specimens [4.7]. There should be little doubt that the interaction curve shown in Figure 4.24 more accurately predicts the strength of a section than does the linear relationship currently used for concrete masonry (also shown in Figure 4.24).

Balanced failure occurs when the compressive strain in the concrete and the tensile strain in the reinforcing reach ultimate and yield levels simultaneously. Whitney concluded that a compressive stress block of constant depth could be used to predict the behavior of concrete members which were expected to have compressive failures. The effective depth of the compressive stress block was set at $0.54d$. Hence, the appropriate component model is as shown in

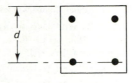

(a) Column Section

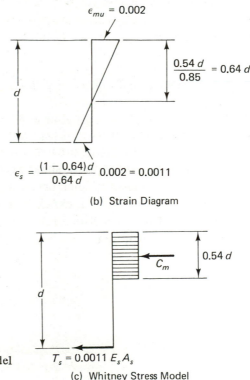

(b) Strain Diagram

Figure 4.25 Whitney stress model for columns

(c) Whitney Stress Model

TABLE 4.3   Equivalent Solid Thickness–Hollow-Unit
Masonry Walls[a] [4.8]

Grout Spacing	Nominal Unit Thickness (in.)			In-Plane Shear Area–8-in. block (in.²/ft)
	6	8	12	
Solid grouted	5.6	7.6	11.6	88.5
Grout at:				
16 in. oc	4.5	5.8	8.5	60.9
24 in. oc	4.1	5.2	7.5	50.5
32 in. oc	3.9	4.9	7.0	45.3
40 in. oc	3.8	4.7	6.7	42.3
48 in. oc	3.7	4.6	6.5	40.5
No grout	3.4	4.0	5.5	

[a]Some special shapes will vary from the figures shown.

Robert R. Schneider and Walter L. Dickey, *Reinforced Masonry Design*, © 1980, p. 189. Reprinted by permission of Prentice-Hall, Inc., Englewood Cliffs, N.J.

Figure 4.25.   A similar development is probably appropriate for concrete masonry.

In the tension control region the model used to define flexural strength (M_u) is always conservative and can be used when axial loads below P_b are anticipated (see Figure 4.24).

4.10 MEMBERS SUBJECTED TO SHEAR

Tests performed to date on wall piers serve as the basis for establishing allowable shear stresses. Shear stresses are determined by dividing the force by the total area resisting the force, mortar area plus grout. Thus, the appropriate shear area for a wall or beam is simply the area of the masonry pier or beam. For walls that are partially grouted, equivalent wall thicknesses are often used as a convenient conversion. Table 4.3 lists equivalent wall thicknesses for various typical levels of grouting.

5

STRENGTH DESIGN

5.1 GENERAL

The design procedures currently employed in masonry structures are based on member force capacities obtained using a small deflection linear elastic theory and axial force, moment and shear demands that are obtained using working loads that are much less than actual extreme loads. This so-called *working stress method* was used in the design of concrete structures until the American Concrete Institute (ACI) introduced an alternative design method in the 1956 edition of the *Building Code Requirements for Reinforced Concrete* based on the principles of strength design. Use of *strength design* in reinforced concrete structures has progressed to the point where the working stress design method is an alternative method relegated to an appendix in the current ACI Code (ACI 318-83) [5.1].

The behavior of reinforced concrete masonry is similar to that of reinforced concrete in many respects. This assumption was accepted by the engineering profession when it accepted working stress design as being acceptable for concrete as well as for concrete masonry. The same rationale that prompted the adoption of a strength design approach in concrete should lead to the adoption of strength design for masonry structures. Why should a new and seemingly more complex theory be proposed for concrete masonry if working stress theory accurately predicts the behavior of concrete and concrete masonry in the range of stresses that normally occur in structures? This very legitimate question must be answered to the satisfaction of the design engineer if the strength method is to be accepted by the engineering profession. Basically, there are two features of strength design that require its adoption. Working stress design, although it can accurately predict member behavior at low stress levels, does not accurately predict the *failure load* nor the *mode of failure* of a member. Strength design does. The inability of a design procedure to accomplish these essential tasks precludes its use as the only design procedure, especially when earthquake loads are a consideration and stresses are expected to be at or near ultimate. Good design practice requires that an engineer develop a building with uniform reliability, and even more important, that the engineer understand what the real strength and post-yield behavior of the components of the structure are so that the engineer can prevent brittle failures that occur during conditions of extreme overload. Only strength design accomplishes these objectives.

A two-level design must be used for concrete masonry and both the working stress theory and the strength theory must be understood by the structural engineer. *Serviceability of a member* when subjected to loads which normally occur once every 20 or 30 years can be evaluated only by using working stress design procedures. The *strength of the member* must be known to design the member for severe environmental loads that can be reasonably expected to occur during the life of the building. Member strength capacity can only be predicted using strength theory.

The structural mechanics of reinforced concrete masonry are presented in Chapter 3 for small and large strains. It shows that there exists a significant difference in the behavior of masonry in each range. The strength of a masonry structural element can be estimated using strength design procedures because it:

1. Recognizes that the relationship between stress and strain is nonlinear

2. Allows the designer to assess the ductility of a structure loaded into the post-yield range of the steel

3. Allows a more rational incorporation of the uncertainty present in design

4. Allows a probabalistic assessment of structural safety

The fundamental goal of strength design is the same as it is for all other methods of design in that it strives to provide a higher level of resistance than is required to support the anticipated loads. This fundamental relationship may be expressed as

(Design capacity or resistance) $\geq$ (effect of design loads)   (5.1)

Thus, the real strength of a masonry member must be at least equal to the task of resisting the effects of the applied design loads. The member capacity commonly referred to as the *nominal capacity* of the member (M_n, V_n, P_n) is the strength calculated using accepted analytical procedures. A strength reduction factor, or ϕ *factor*, is used to modify the nominal capacity to obtain the strength of the member for design purposes.

Factors affecting the value of ϕ include uncertainties in workmanship, difference between design loads and maximum loads expected during the life of the structure, material strength, physical dimensions, and failure mode. The last consideration is very important because it is usually desirable to provide greater reliability where dealing with brittle modes of failure. The collapse of a column, occurring as the result of the brittle compressive failure of the masonry, is less desirable than the ductile yielding of the tension steel in an under-reinforced member. Thus, it is reasonable that a smaller strength reduction factor (i.e., more conservative) be used for columns than for beams. Chapter 6 and Appendix A discuss the development of the ϕ factors for various situations encountered in design.

The design expressions resulting from Equation (5.1) are

$$\phi M_n \geq M_u \qquad (5.2a)$$

$$\phi P_n \geq P_u \qquad (5.2b)$$

$$\phi V_n \geq V_u \qquad (5.2c)$$

where ϕ = strength reduction factor (less than unity)

M_n, P_n, V_n = nominal (assumed) strength of member for moment, axial, or shear capacity, respectively, as computed by prescribed equations

M_u, P_u, V_u = required ultimate strength

The required ultimate strength is a function of the design loads which are obtained by multiplying nominal 50-year loads by a load factor. Load factors reflect the uncertainty associated with various types of loading combinations and are intended to provide a *uniform* reliability for structural components. The development of these load factors, as well as the philosophy behind them, are presented in Chapter 6.

A recent study [5.2] has proposed to the American National Standards Institute (ANSI) the following load factors and loading combinations for the design of building structures:

$$1.4D \tag{5.3a}$$

$$1.2D + 1.6L \tag{5.3b}$$

$$1.2D + 1.6S + (0.5L \text{ or } 0.8W) \tag{5.3c}$$

$$1.2D + 1.3W + 0.5L \tag{5.3d}$$

$$1.2D + 1.5E + (0.2S \text{ or } 0.5L) \tag{5.3e}$$

$$0.9D - (1.3W \text{ or } 1.5E) \tag{5.3f}$$

where D, L, S, W, and E are dead, live, snow, wind, and earthquake service loads or forces, respectively.

Equation (5.3) combines dead loads with amplified values of the remaining load types. It is thought that the proposed load criterion is a better model of how loads actually combine during the life of a structure than are the current load factors used in the ACI Code [e.g., $1.4D + 1.7L$ for Equation (5.3b)] or AISC Specification [e.g., $1.7D + 1.7L$ for Equation (5.3b)]. These load combinations also seem consistent with observations related to the failure of buildings when one load was increased to an extreme value [5.3]. The additive combinations of load in Equations (5.3) maintain the same effect of the factored dead load except in Equation (5.3a), which attempts to ensure a minimum level of safety where the dead load is dominant. Equation (5.3f) recognizes the fact that all dead loads may not be acting on the structure when it is called on to resist the load reversals that occur during seismic or wind loadings.

5.2 STRENGTH DESIGN OF FLEXURAL ELEMENTS

The structural mechanics describing the flexural behavior of reinforced concrete masonry are presented in Section 3.3 and describe
three failure mechanisms for flexural elements. If the tension steel
yields before the masonry crushes, the member is considered under-
reinforced and undergoes a tension failure. This failure mechanism
is preferable to the other two mechanisms because the distress in the
member is well announced by considerable cracking in the masonry
member prior to failure. A balanced failure occurs when the steel
starts yielding at the same time the masonry crushes. An over-
reinforced section undergoes a compression failure because the
masonry crushes before the steel yields. It is important to prevent
these last two modes of brittle failure, and this may be accomplished
by limiting the amount of steel to an amount less than that which
will cause a balanced failure.

The steel ratio, ρ, is used to express the ratio of the area of the
tension steel to the gross cross-sectional areas of the member:

$$\rho = \frac{A_s}{bd} \tag{5.4}$$

At balanced failure, the steel ratio is denoted ρ_b. This balanced
steel ratio may be obtained by assuming that the strain in the steel
is equal to its yield strain, ϵ_y, and the strain level in the masonry
has reached its ultimate value, ϵ_{mu}. Referring to Figure 5.1, and

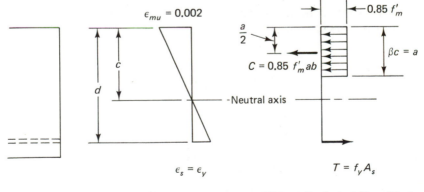

Figure 5.1 Balanced failure condition

through the use of similar triangles, an expression for ρ_b may be obtained. *If the ultimate strain in the masonry is assumed here to be $\epsilon_{mu} = 0.002$*, as discussed in Section 3.2, the balanced steel ratio is

$$\rho_b = \frac{A_{s\,bal}}{bd} = \frac{0.85 f'_m}{f_y} \beta \left(\frac{0.002}{0.002 + f_y/E_s} \right) \tag{5.5}$$

where $A_{s\,bal}$ denotes the area of steel at balanced failure.

If $E_s = 29{,}000{,}000$ psi and $\beta = 0.85$ (refer to Section 3.3 for a discussion of β) are substituted into Equation (5.5), the balanced steel ratio is

$$\rho_b = \frac{0.72 f'_m}{f_y} \left(\frac{58{,}000}{58{,}000 + f_y} \right) \tag{5.6}$$

Thus, balanced failure occurs when $\rho = \rho_b$. A tension failure in an under-reinforced section occurs when $\rho < \rho_b$, and a compression failure in an over-reinforced section occurs when $\rho > \rho_b$. The ACI code requires that ρ not exceed $0.75\rho_b$ to ensure that a tension failure will occur, because the actual yield strength of the steel exceeds that specified by as much as 33 percent. This recommendation seems applicable to strength design in masonry also, and as a result, the reinforcing steel permitted in a flexural member should be limited to

$$\rho_{max} \leqslant 0.75\rho_b \qquad \text{or} \qquad \frac{0.54 f'_m}{f_y} \left(\frac{58{,}000}{58{,}000 + f_y} \right) \tag{5.7}$$

The nominal moment capacity of the member, M_n, may be obtained from the discussion of mechanics in Section 3.3 and is given by

$$M_n = A_s f_y \left(d - \frac{a}{2} \right) \tag{5.8}$$

where

$$a = \frac{A_s f_y}{0.85 f'_m b} = \rho \left(\frac{f_y d}{0.85 f'_m} \right)$$

Substituting the foregoing value of a into Equation (5.8) and performing some algebraic simplifications, the expression for M_n becomes

$$M_n = \rho f_y bd^2 \left[1 - 0.59 \left(\frac{\rho f_y}{f'_m}\right)\right] \qquad (5.9)$$

Insofar as the design of a flexural member is concerned, the nominal moment capacity of the member reduced by the ϕ factor must, as previously indicated, be greater than or equal to the required moment capacity produced by the factored loads:

$$\phi M_n \geqslant M_u \qquad (5.10)$$

When, for given values of b and d, the required moment demand is greater than that which can be resisted by the available compression region of the masonry, compression reinforcing must be added. The addition of steel in the compression region can quite often increase the capacity of the member without increasing its physical size or altering the mode of failure.

The mechanics of a doubly reinforced member may be modeled as shown in Figure 5.2 such that a portion of the design moment is resisted by the moment couple, consisting of the compression force in the masonry and the force in a portion of the tension steel, and the remainder of the design moment is resisted by the moment couple between the force in the compression steel and the remaining tension steel. The maximum design moment that can be carried by a singly reinforced concrete masonry member (without any compression steel), ϕM_{nm}, is a function of the balanced steel ratio and is

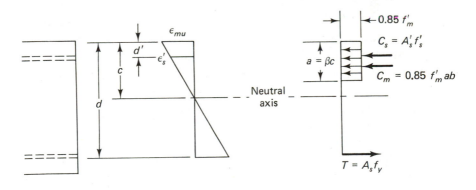

Strain   Whitney Equivalent Stress Block

Figure 5.2 Masonry member reinforced with compression steel

given by

$$\phi M_{nm} = \phi \rho_{\max} f_y bd^2 \left[1 - 0.59 \left(\frac{\rho_{\max} f_y}{f_m'}\right)\right] \qquad (5.11)$$

where $\rho_{\max}$ is as developed in Equation (5.7). Thus, the contribution to the flexural capacity provided by the compression reinforcement, $\phi M_{nm}'$, is the difference between the required ultimate moment strength, M_u, and that which is provided by the masonry acting alone as a singly reinforced member:

$$\phi M_{nm}' = M_u - \phi M_{nm} \qquad (5.12)$$

If it is assumed that the compression steel yields (the stress in the compression steel is $f_s' = f_y$), the ratio of compression steel, ρ', required to provide $\phi M_{nm}'$ is

$$\rho' = \frac{A_s'}{bd} = \frac{\phi M_{nm}}{f_y(d - d')\,bd} \qquad (5.13)$$

The total amount of tension steel, $A_{s\,\text{tot}}$, required for internal equilibrium is

$$A_{s\,\text{tot}} = \rho bd = (0.75\rho_b + \rho')\,bd \qquad (5.14)$$

The first term on the right-hand side of Equation (5.14) represents the tension steel required to balance the compressive force in the masonry, and the second term is the tension steel required to balance the compression force carried by the compression steel provided that the compression steel has yielded. Only that portion of the tension steel balanced by the compression in the masonry need be reduced by the 0.75 factor. This is because the additional tension steel is exactly balanced by the compression reinforcement and each presumably has the same ratio of actual yield strength to specified yield strength.

An alternative expression for the reduced nominal capacity of a doubly reinforced member if the compression steel yields is

$$\phi M_n = \phi \left[(A_s - A_s')\,f_y \left(d - \frac{a}{2}\right) + A_s' f_y(d - d')\right] \geqslant M_u \quad (5.15)$$

where

$$a = \frac{(A_s - A_s')\,f_y}{0.85\,f_m'\,b}$$

The assumption that the compression steel has yielded may be checked by assuring that

$$\rho - \rho' \geqslant \frac{0.85 f'_m \beta d'}{f_y d} \left[\frac{0.002 E_s}{(0.002 E_s) - f_y} \right]$$

$$\geqslant \frac{42,000 f'_m d'}{f_y d (58,000 - f_y)}$$

(5.16)

(The check described above assumes an ultimate strain of E_{mu} = 0.002. Other values could be assumed as required.)

Of course, if the compression steel has not yielded, Equation (5.15) will overestimate the strength of the section. The stress in the compression steel, f'_s, must then be obtained by using the strain diagram in Figure 5.2, and is

$$f'_s = \epsilon'_s E_s = 0.002 \left(\frac{a - \beta d'}{a} \right) E_s$$

$$= 58,000 \left(\frac{a - 0.85 d'}{a} \right)$$

(5.17)

The expression equivalent to Equation (5.15) for the reduced nominal moment capacity is

$$\phi M_n = \phi \left[0.85 f'_m ab \left(d - \frac{a}{2} \right) + A'_s f'_s (d - d') \right] \geqslant M_u \quad (5.18)$$

Doubly reinforced sections in masonry where the compression steel is required for strength are fairly rare, but there are occasions where compression reinforcement is added to control long-term deflections or, more frequently, provide for tension resistance in two directions because of anticipated load reversals during earthquakes. When compression steel is added to the section, the strength characteristics should be investigated using the expressions in this section to ensure that an accurate estimate of the yield strength of the member is obtained, for only then can the designer understand and control the mode of failure.

5.3 STRENGTH DESIGN FOR FLEXURE AND AXIAL LOADS IN MASONRY WALLS AND COLUMNS

It was suggested in Section 4.3 that an appropriate method to model the capacity of a member subjected to both bending and axial loads is an interaction approach which accounts for the relationship between the stresses caused by bending and axial loads. An *interaction diagram*, such as that shown in Figure 5.3, may be constructed by establishing the capacity of the member under various combinations of axial and flexural loads. Although an infinite number of points may be calculated, the critical points identified by numbers 1 through 6a on Figure 5.3 should be more than sufficient to construct an accurate interaction diagram. Each point is described by the axial capacity P_n and moment capacity M_n available for a given combination of P_u and M_u. Thus, M_n can be computed for a given P_n, or vice versa.

At one extreme, point 1, when no externally applied moment is imposed on the wall the axial capacity of the wall, P_0, is the compressive capacity of the masonry plus reinforcing steel (including the small amount of masonry displaced by the steel):

$$\phi P_n = \phi P_0 = \phi(0.85 f'_m tb + A_s f_y) \tag{5.19}$$

The other extreme, point 6, is where the capacity of the member is the nominal flexural capacity of the wall and no externally applied axial load is imposed on the wall:

$$\phi M_n = \phi 0.85 f'_m ab \left(d - \frac{a}{2}\right) \tag{5.20}$$

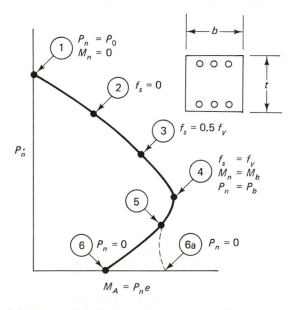

Figure 5.3 Interaction diagram of an eccentrically loaded member

The intermediate points may be established by choosing several conditions of strain and, from the force equilibrium and stress-strain relationships developed in Chapter 3, calculating P_n and M_n. The intermediate points chosen for investigation are usually: $f_s = 0$, $f_s = 0.5f_y$, and $f_s = f_y$. The last point (point 5 in Figure 5.3), $f_s = f_y$, is the location that corresponds to the balanced conditions, P_b, and M_b. Following current practice in the design of concrete columns, the value of ϕ is allowed to increase, in a linear manner, from the value used in the design of a column to the value used in the design of a flexural member when the axial load, P, is less than $0.1 f'_m A_g$ or ϕP_b, whichever is less. This recognizes that the failure of the member is controlled by flexure, and at this load level can exhibit ductile behavior. The increased capacity is shown on Figure 5.3 at point 6a.

The load-induced moment on a wall is a function of lateral wall deflection. If the wall is slender, the lateral deflection can produce moments that are significant relative to the moment obtained using small deflection theory. What is slender? To best understand this, consider the following development.

Figure 5.4 shows the forces acting on a simply-supported slender wall. The summation of moments about the bottom of the wall gives

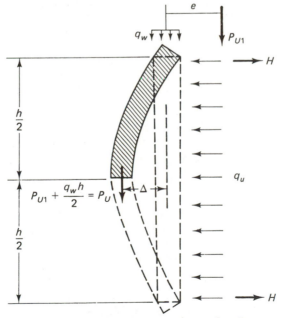

Figure 5.4 Loading geometry of wall

the equation for the horizontal force at the upper wall support. That is,

$$P_{u1}e + Hh - \frac{q_u h^2}{2} - q_w h(\Delta_a) = 0 \qquad (5.21)$$

where P_{u1} = vertical load on wall

$\quad e$ = eccentricity of vertical load

$\quad q_u$ = lateral load on wall per linear foot

$\quad q_w$ = weight of wall per linear foot

$\quad \Delta_a$ = "effective" lateral deflection used to estimate dead load moment

If we assume that

$$\Delta_a = \frac{2\Delta}{3} \qquad (5.22)$$

where Δ is the wall's midheight lateral deflection, then

$$H = \frac{q_u h}{2} - \frac{P_{u1}e}{h} + \frac{2q_w \Delta}{3} \qquad (5.23)$$

The first term corresponds to the classical small deflection reaction, the second term represents the change in the magnitude of the force due to an eccentric wall loading, and the third term incorporates the lateral wall deflection.

If we take the moment about the midheight of the wall, the moment induced on the cross section from the external loads is

$$M = H\left(\frac{h}{2}\right) + P_{u1}(\Delta + e) + \left(\frac{q_w h}{2}\right)\Delta_b - \left(\frac{q_u h}{2}\right)\frac{h}{4} \qquad (5.24)$$

where Δ_b is the "effective" lateral deflection used to estimate dead load moment. If we assume that

$$\Delta_b = \frac{\Delta}{3} \qquad (5.25)$$

which is consistent with Δ_a above and substitute H into the moment equation, it follows that

$$M = \frac{q_u h^2}{8} + \frac{P_{u1}e}{2} + \left(P_{u1} + \frac{q_w h}{2}\right)\Delta \qquad (5.26)$$

The first term corresponds to the moment due to the classical small deflection moment from the lateral load, the second term corresponds to the moment due to the eccentric vertical load on the wall, and the third term represents the moment due to large lateral deflections. This last term can be referred to as the $P - \Delta$ *load*. When a wall is slender and Δ increases in magnitude, this term increases in value.

The moment M and lateral force H are a function of Δ, which in turn is a function of the wall's cross-sectional properties and steel reinforcement as well as the moment M and the lateral load H. Therefore, the problem of calculating the moment M is iterative, and fortunately it is one that has been shown to converge rapidly.

Current working stress design methods for *slender masonry walls* incorporate a stress reduction approach. The maximum allowable stress is reduced by a nonlinear term which is controlled by the height-to-thickness (h/t) ratio of the wall. Consideration of fixity conditions at the base and top of the wall may be incorporated by computing the effective height of the wall as discussed in Section 4.3. Research indicates that walls with an h/t less than 10 do not exhibit stability problems associated with slenderness [5.4]. The current code equation allows the use of 98 percent of the maximum allowable stress with an h/t ratio of 10, so it seems that slenderness is not perceived as a major problem at least at this level.

There are three methods which are available for the strength design of concrete walls [5.5], all of which consider the effects of combined loads. Recognizing the similarities between the behavior of concrete masonry walls and concrete walls, it seems reasonable to adapt these approaches to the strength design of masonry walls. One method places limiting restrictions on the loading and dimensions of the wall and basically is empirical in approach. The second method is a so-called "rational" approach which checks an assumed trial wall section for axial and flexural capacity when subjected to loading that incorporates the $P\text{-}\Delta$ loading effect. The third method uses a moment magnifier to account for the secondary bending moments due to slenderness.

The *empirical approach* can be used provided that the limitations under which the method can be applied are met. Bearing masonry walls loaded within the middle third of its thickness (eccentricity no greater than $t/6$) with an h/t ratio of 25 or less may be designed using this empirical approach. Additional transverse moments caused by out-of-plane wind or earthquake loads are not explicitly considered in determining the axial capacity of the wall; however, these moments are used to determine the total eccentricity of the applied load to see if the resulting load still falls within the middle third.

The nominal axial load strength of a wall is assumed to equal

$$P_n = 0.55 f'_m A_g \left[1 - \left(\frac{kh}{32t}\right)^2\right] \qquad (5.27)$$

where A_g = gross cross-sectional bearing area of wall under concentrated load (or area of wall per unit length under line load)

h = actual height of wall between vertical supports

k = effective length factor: walls braced top and bottom against lateral translation and (1) restrained against rotation at one or both ends (top and/or bottom): 0.8; (2) unrestrained against rotation at both ends: 1.0; and (3) walls not braced against lateral translation: 2.0

Therefore, the ultimate axial load computed using the factored axial forces must be less than the evaluated nominal capacity:

$$\phi P_n \geqslant P_u \qquad (5.28)$$

For walls that do not meet the limitations imposed on Equation (5.27), methods 2 and 3 are available and require the wall to have a capacity equal to the sum of the superimposed factored axial dead and live loads, P_{u1}, factored wall dead load for the upper one-half, P_{u2}, along with the factored lateral load from wind or earthquake, q_u (Figure 5.4). The moment capacity of a wall section is calculated, assuming that axial strength does not govern the design, and it is checked against the moment generated under the applied lateral load and by the P-Δ effect [5.5].

Although most walls are loaded at a level which is considerably less than their axial load strength, a check can be made to determine if flexure controls the design, that is,

$$\phi P_b \geqslant P_u \qquad (5.29)$$

in which

$$P_b = 0.85 f'_m b a_b - A_s f_y$$

where

$$a_b = \left(\frac{58,000}{58,000 + f_y}\right) \beta d$$

The nominal moment capacity of the assumed wall section loaded with a concentrically applied load may be determined from force and moment equilibrium (see Figures 5.5 and 5.6). The axial load is

$$P_u = C - T$$

or

$$0.85 f'_m ba = P_u + A_s f_y \tag{5.30}$$

and solving for a yields

$$a = \frac{P_u + A_s f_y}{0.85 f'_m b} \tag{5.31}$$

Summing the internal and external moments yields

$$M_u + P_u \left(d - \frac{t}{2}\right) - C\left(d - \frac{a}{2}\right) = 0$$

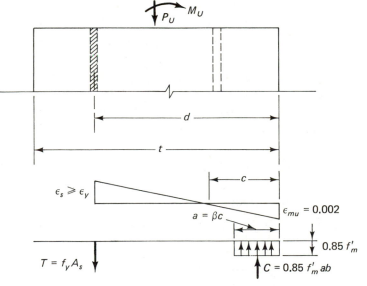

Figure 5.5 Steel at two faces (ignoring compression steel)

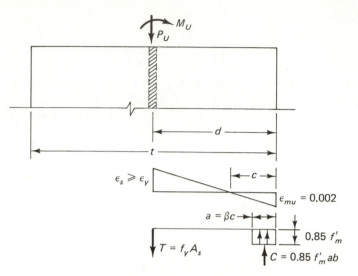

Figure 5.6 Steel at center of wall

Substituting Equation (5.30) for C, and assuming $M_n = M_u$, the nominal moment capacity of a member with steel at two faces is

$$M_n = (P_n + A_s f_y) \left(d - \frac{a}{2}\right) - P_n \left(d - \frac{t}{2}\right) \qquad (5.32)$$

In the more typical case with steel in one layer of reinforcement at the centerline of the wall, the nominal moment capacity is

$$M_n = (P_n + A_s f_y) \left(d - \frac{a}{2}\right) \qquad (5.33)$$

The maximum out-of-plane deflection, Δ, from the combined loads may be approximated as

$$\Delta \simeq \frac{5M_n h^2}{48 E_m I_e} = \frac{M_n h^2}{9.6 E_m I_e} \qquad (5.34)$$

where I_e denotes the effective moment of intertia.

The current of ACI Code recommends an expression for the effective moment of inertia which accounts for the cracking attributable to the applied moment in a *flexural* member. An expression for the effective moment of inertia in a masonry member under combined loadings has yet to be derived and experimentally verified, but a conservative approximation may be obtained by assuming a

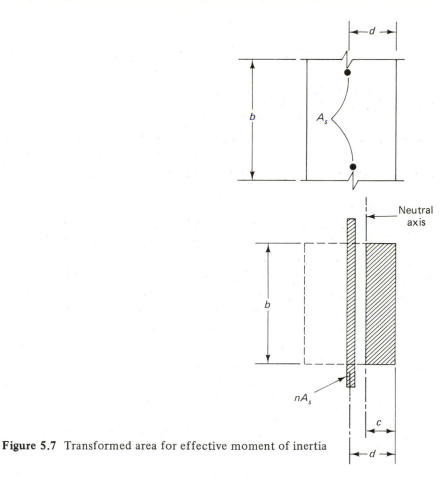

Figure 5.7 Transformed area for effective moment of inertia

transformed cracked section as shown in Figure 5.7. Thus, the moment of inertia of the transformed, cracked section, I_t, is

$$I_t = I_s + I_m \qquad (5.35)$$

where I_s and I_m represent the moment of inertia of the steel and the masonry in the compression region, respectively. Thus, about the neutral axis,

$$I_s = nA_{\text{eff}}(d - c)^2 \qquad (5.36)$$

where $A_{\text{eff}} = (P_n + A_s f_y)/f_y$

$n = E_s/E_m$

c = distance of neutral axis to extreme compression fiber

and

$$I_m = \frac{bc^3}{12} + cb\left(\frac{c}{2}\right)^2 = \frac{bc^3}{3} \tag{5.37}$$

where b is the effective width of the wall (refer to Section 3.3).

If the imposed moment, M_u, is less than the reduced moment capacity, ϕM_n, the wall section is acceptable.

$$\phi M_n \geqslant M_u \tag{5.38}$$

This may be determined by comparing Equation (5.26) with Equation (5.32) or (5.33), multiplied by the appropriate ϕ factor.

This method is limited in terms of accuracy in that it assumes that the wall is simply supported when computing the secondary moment M. (The value of the moment caused by the applied lateral load may be computed using the appropriate model for the wall as shown in Section 4.3.) Another approximation of the secondary moment may be obtained if a moment magnifier approach is used [5.1].

The moment magnifier attempts to account for secondary bending moments caused by P-Δ effects by increasing the ultimate design moment M_u by a coefficient, δ. The magnified moment concept is illustrated on the interaction diagram in Figure 5.8, and is similar to

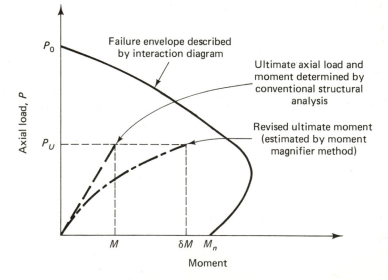

Figure 5.8 Moment magnification in a slender member

the approach developed by the AISC and adapted to concrete walls and columns by the ACI. The value of δ is given by

$$\delta = \left(\frac{C_m}{1 - P_u/\phi P_c} \right) \geq 1.0 \qquad (5.39)$$

where $C_m = 0.6 + 0.4(M_1/M_2) \geq 0.4$

   $P_c = \pi^2 EI/(kh)^2$

   $M_{1,2}$ = smaller and larger moments at wall ends (i.e., top and bottom)

The parameter C_m represents an end-effect factor where M_1 is the smaller of the ultimate moments at the ends of the wall and the ratio is assumed to be positive if the wall is bent in single curvature and negative if bent in double curvature. Inasmuch as the wall is not usually fixed at the top in single-story walls, the value of C_m often defaults to 0.6.

Determination of the moment of inertia again presents a problem in that the section is often cracked, but the cracking varies with the intensity of the applied moment and axial load. A reasonably conservative estimate of EI may be obtained by

$$EI = 0.4 E_m I_g \qquad (5.40)$$

which ignores the contribution of the steel in lightly reinforced walls. Future research in masonry may suggest that a creep factor be incorporated to account for the reduction in stiffness due to sustained loads.

In the development of the moment magnifier approach in both concrete and steel, the effective height factor k takes into account the degree of lateral and rotational stiffness at the ends of the wall. The values of k are the same as those presented in Table 4.2.

Most slender concrete masonry walls used in one- to four-story buildings support only relatively small axial loads and, as a result, the importance of eccentrically applied loads and P-Δ effects is often overcompensated for.

In 1981, the Structural Engineers Association of Southern California (SEAOSC) tested 32 slender concrete, brick, and concrete masonry panels subjected to a constant axial and increasing lateral load [5.6]. Panel capacities were predicted using the strength method developed by SEAOSC [5.7]. The procedure for calculating ultimate

moments and deflections is presented in Equations (5.33) and (5.34).
Load deflection results of these tests are presented in Figure 5.9.
A close correlation was obtained between calculations and test data.

The design of columns and pilasters is similar to the design of
axially loaded walls. The same concept using an interaction diagram
presented in Section 5.3 is used to establish the capacity of a pilaster
or column with a combined loading.

The axial strength, P_n, of an axially loaded column section with
no applied moment is assumed to be the sum of the masonry and
steel strengths:

$$P_n = 0.85 f'_m A_m + A_s f_y \qquad (5.41)$$

where A_m is the area of the masonry.

The design relationship that must be satisfied is

$$\phi P_n \geqslant P_u \qquad (5.42)$$

Combined loadings in columns are not as common in masonry
as they are in concrete because the continuity that is inherent in
reinforced concrete structures is difficult to achieve in masonry.
Most loading conditions that cause both bending and axial loads are
a result of eccentrically applied beam reactions or lateral loads imposed during earthquakes. The strength of a concrete section loaded
in flexure and compression is established from an interaction diagram which defines the relationship between the axial stresses due to
compression and stresses caused by bending. The same type of force
equilibrium relationship used to develop the interaction diagram for
a wall in Section 5.3 may be used to establish the capacity of a
column in flexure and compression.

In many cases, the physical dimensions of a column are dictated
by basic concrete block sizes or other architectural considerations,
and frequently have structural capacities far in excess of those
required to support the anticipated loads. As a result, minimum steel
requirements for columns often require an arbitrarily large amount
of steel. The current building codes permit the effective area of the
column section to be reduced by up to 50 percent, when computing
the minimum steel, if the reduced section can be shown to be capable of supporting the imposed loads.

Slenderness effects are provided for in columns in the same way
as they are in walls. The previously presented methods using a so-
called "rational" approach or a "moment magnifier" method to

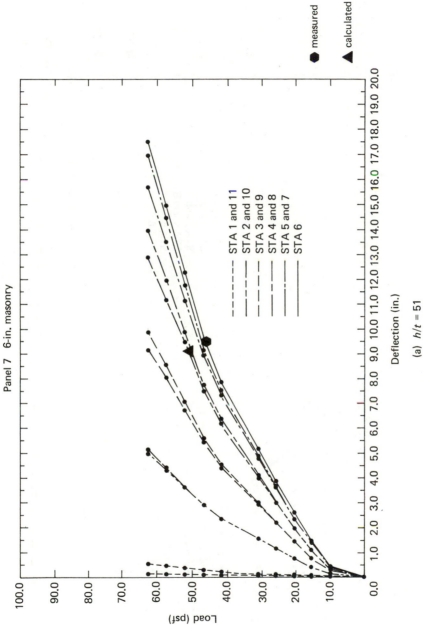

Panel 7   6-in. masonry

◆ measured

▲ calculated

STA 1 and 11
STA 2 and 10
STA 3 and 9
STA 4 and 8
STA 5 and 7
STA 6

Deflection (in.)

Load (psf)

(a) *h/t* = 51

Figure 5.9 Load/deflection curves (slender walls)

173

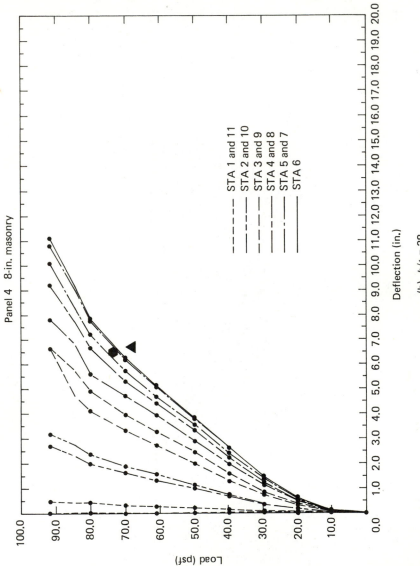

Panel 4   8-in. masonry

STA 1 and 11
STA 2 and 10
STA 3 and 9
STA 4 and 8
STA 5 and 7
STA 6

Load (psf)

Deflection (in.)

(b) $h/t = 38$

⬢ measured

▲ calculated

Figure 5.9 (*Continued*)

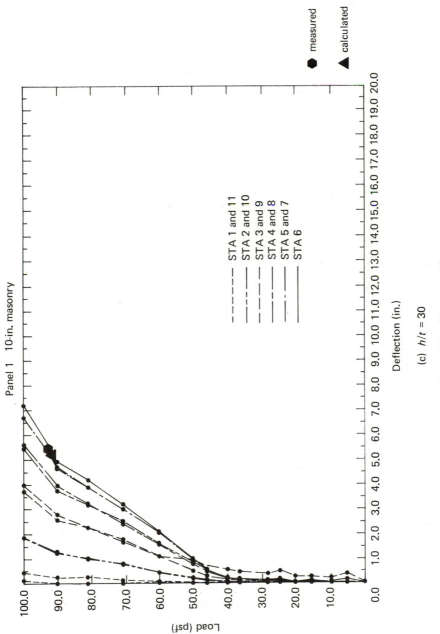

(c) $h/t = 30$

Figure 5.9 (*Continued*)

account for the secondary moments caused by slenderness considerations may also be used to calculate the capacity of a column. Slenderness is usually not a controlling influence in column design, and procedures developed by Whitney (as discussed in Section 4.9) may be used to model column sections in order to rapidly develop axial and flexural capacities of the column sections.

Pilaster design is identical to column design in most respects. The axial and moment capacities may be established in exactly the same manner as in the design of walls. Pilasters supporting axial and bending loads are quite common, because pilasters support gravity loads and provide lateral support for walls.

5.4 STRENGTH DESIGN OF MEMBERS SUBJECTED TO SHEAR

The failure of a concrete masonry member in shear generally represents an undesirable mode of failure because of the suddenness associated with this type of failure. An understanding of shear-related phenomona allows the designer to provide a sufficient amount of reinforcing steel in the proper locations to ensure satisfactory structural behavior under both service and extreme-type loadings. Section 3.5 presents an explanation of the mechanics of shear in concrete masonry members and the influence of member geometry on the failure modes expected in structural elements loaded in shear. In this section we discuss some of the shear design provisions contained in the existing concrete building codes as they relate to concrete masonry, recent test results concerned with the structural performance of concrete masonry shear walls, and design considerations for shear walls.

The shear stress in all members should be checked; however, in very long beams where the shear span-to-depth ratio, a/d, is greater than or equal to 6, it is unusual to find that shear is the controlling mode of failure. This occurs because the limitation on the longitudinal tension steel generally results in a flexural strength in the beam which is less than the shear strength, and hence member strength will be limited by yielding of the tension steel.

Flexural members with a/d ratios less than or equal to 6 should always be investigated for shear. It is assumed that the diagonal tension stress producing the cracking in a concrete masonry member is approximately equal to the shear strength of the masonry acting alone, V_m. Any required additional shear strength must be provided by the shear reinforcing except for members with shear spans (a/d) less than 1 because of the additional shear strength developed by arch action. The total shear strength of the structural element is given by the sum of the shear strength of the concrete masonry and that of shear reinforcement:

$$V_n = V_m + V_s \qquad (5.43)$$

Current design provisions for shear in concrete masonry, as defined in the UBC [5.8], do not recognize the concept presented in Equation (5.43). Instead, it requires that if the required shear strength of the member exceeds the strength of the masonry (V_m), shear reinforcement must be designed to carry the total shear load $(V_n = V_s)$. Where anticipated member deflections are large and substantial cracking in the concrete masonry is anticipated, the addition of shear steel designed to carry the total shear load may be reasonable. However, experimental data gathered over the last two decades on both unreinforced and reinforced masonry members subjected to static and cyclic shear tests suggest that this is an extremely conservative approach to design. A more reasonable approach is to require a minimum amount of shear reinforcement in a member when the shear stress is greater than or equal to some predetermined percentage of the shear strength of the concrete masonry. This concept has been adopted by the ACI in their design recommendations for shear in reinforced concrete. The shear reinforcing provided in a concrete masonry member must be designed to reduce the possibility of a brittle failure in shear. The minimum quantity should be predicated on preventing a brittle failure at the formation of the first diagonal tension crack. To accomplish this, the shear reinforcing should behave elastically when the cracking load is reached. Shear reinforcing should also be *limited* if ductile behavior is desired. The truss analogy discussed in Chapter 3 requires that a diagonal concrete masonry compression strut form. The compressive capacity of this strut should be greater than the corresponding yield load on the tension strut.

The ACI specifies a minimum area of shear steel, A_{v_c}, in a reinforced concrete member if $V_u \geqslant \phi V_c/2$ and it is equal to

$$A_{v_c} = 50\left(\frac{bs}{f_y}\right) \qquad (5.44)$$

where s = spacing of the shear steel

b = thickness of member

The formulation of Equation (5.44) is based on an assumption that the shear strength of concrete is approximately 100 psi for 3000-psi concrete. The current ACI provisions required shear steel if the shear stress is greater than or equal to one-half of the concrete shear strength. This minimum shear steel requirement then provides the member with an amount of shear steel such that it will not fail immediately in diagonal tension upon reaching the inclined cracking strength in the concrete. Minimum shear reinforcing in concrete masonry should be consistent with that used for concrete beams. A proportional relationship between the minimum shear steel in concrete and concrete masonry may be obtained. If it is assumed that $V_m = 1.7\sqrt{f'_m}$ for $f'_m = 1500$ psi [5.9] and $V_c = 2.0\sqrt{f'_m}$ for $f'_c = 3000$ psi [5.1], then the minimum area of shear steel, $A_{v\,min}$, is

$$
\begin{aligned}
A_{v\,min} &= \frac{V_m}{V_c}\left(50\,\frac{bs}{f_y}\right) \\[2mm]
&= \frac{1.7\sqrt{f'_m}}{2\sqrt{f'_c}}\left(50\,\frac{bs}{f_y}\right) \\[2mm]
&= \frac{1.7\sqrt{1500}}{2\sqrt{3000}}\left(50\,\frac{bs}{f_y}\right) \\[2mm]
&= 30\,\frac{bs}{f_y}
\end{aligned}
\tag{5.45}
$$

Not only is it important that the member be capable of reaching a minimum shear strength to preclude a sudden diagonal tension failure, but the undesirable compressive failure of the diagonal compression struts in the truss mechanism must also be prevented. Research [5.10] also suggests that a biaxial state of strain exists in the web of a member with shear steel because the stirrups through the web transmit tension to the struts through bond. The action of the transverse tensile strain reduces the compressive strength of the concrete masonry. A significant increase in the diagonal compression stress also results when the diagonal compression struts are inclined at an angle less than 45° to the horizontal. There appears to be substantial evidence to limit the diagonal compressive stress to a value significantly smaller than the compressive strength of the concrete masonry.

The ACI controls the diagonal compressive stress by limiting the maximum amount of shear steel in a member such that $V_s \leqslant$

$8\sqrt{f_c'}\,bd$. A similar condition should control the maximum contribution of the truss mechanism for shear resistance in concrete masonry. Using the ratios of the compressive strengths of the concrete masonry (f_m') and concrete (f_c') and assuming that $f_m' = 1500$ psi and $f_c' = 3000$ psi, we have

$$V_s = 8 \left(\frac{f_m'}{f_c'}\right) \sqrt{f_m'}\,bd$$

$$= 8 \left(\frac{1500}{3000}\right) \sqrt{f_m'}\,bd \qquad (5.46)$$

$$= 4\sqrt{f_m'}\,bd$$

If it is assumed that minimum shear steel must be provided if the required shear strength, V_u, is greater than or equal to one-half of the concrete masonry shear strength $V_m/2$, and that the truss analogy operates if $V_u \geq V_m$, a diagram such as that in Figure 5.10 illustrates the three principal ranges to be considered in the design of shear steel.

Within the limits addressed above, the shear reinforcing must carry the difference between the required shear strength and the shear strength of the concrete masonry, $V_u - \phi V_m$. The spacing of

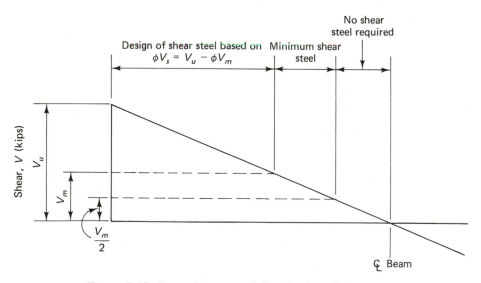

Figure 5.10 Shear diagram and distribution of shear steel

the shear steel, s, is given by

$$s = \frac{\phi A_v f_y d}{V_u - \phi V_m}$$ (5.47)

Recent research has addressed the behavior of concrete masonry shear walls under cyclic loads to establish the predominant mode of failure as well as the important design parameters affecting the behavior of these shear walls. An investigation of a series of reinforced concrete block walls subjected to cyclic loading was conducted at the University of California, Berkeley [5.11]. They examined walls with different height-to-length ratios and varied the amount of vertical and horizontal steel to determine its influence on the performance of the walls.

Figure 5.11 shows the hysteritic behavior of three walls with the same vertical steel ratio, $\rho_v = 0.0017$. Figure 5.11a represents the force/displacement relationship for a wall with no horizontal reinforcing, $\rho_h = 0.00$. The remaining two walls (Figure 5.11b and c) have horizontal steel ratios of 0.0008 and 0.0034, respectively. Although the ultimate strength is not improved appreciably by the addition of the horizontal steel, it can be seen that as the amount of horizontal steel is increased the behavior of the wall is improved. This is demonstrated in two ways. First, the wall is able to undergo greater displacements without losing its load-carrying capacity. Second, it takes the hysteritic loops a longer time to "lie down" under the cyclic loads. This represents less strength degradation for the same number of cycles of load. The shear force and shear stress versus displacement for the three walls discussed is plotted in Figure 5.12.

Figure 3.17 shows the propagation of cracks in the wall with the force/displacement curve shown in Figure 5.11b. The initial cracking occurs in the horizontal bedding joints of the wall (see Figure 3.17), and occurred when the wall was subjected to its highest load (ultimate shear). Once cracking has occurred in the bedding joint the shear capacity of the wall is reduced and a different mechanism is developed to carry shear loads. The area between the force/displacement curves for wall 1 and wall 2 in Figure 5.12 clearly indicates that the horizontal steel is participating in resisting the shear, probably through the truss mechanism. The shear failure of these walls (see Figure 3.17 for wall 2) was accompanied by very

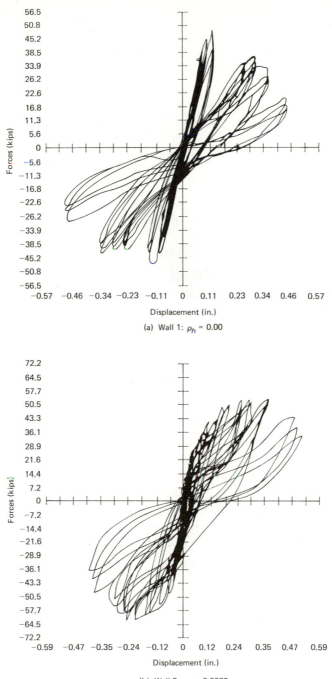

(a) Wall 1: $\rho_h = 0.00$

(b) Wall 2: $\rho_h = 0.0008$

Figure 5.11 Force deflection diagrams for walls with varying amounts of horizontal reinforcement [5.11]

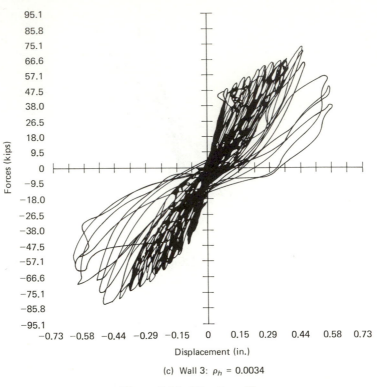

(c) Wall 3: ρ_h = 0.0034

Figure 5.11 (*Continued*)

little flexural cracking, evidenced by the lack of horizontal cracking in anything but the bed joints.

Clearly, the ductility available in walls 2 and 3 (Figure 5.11) is greatly improved over that available in wall 1. If horizontal cracking is assumed to occur at a displacement of about 0.12 in., wall 1 could sustain a displacement of about 0.35 to 0.40 in. without a significant loss of strength. This indicates a displacement ductility of about 3. In walls 2 and 3, the strength of the pier continues to increase and ductilities of between 4 and 5 are available. The horizontal reinforcing provided in wall 2 is the minimum horizontal reinforcing currently required by the UBC [5.8]. Whereas Figure 5.12 seems to indicate a significant improvement in behavior for wall 3 over wall 2, the hysterisis loops of Figure 5.11b and c do not support this contention. It appears, at least for walls with h/d ratios below 1.0, that the inclusion of large amounts of horizontal reinforcing adds little to the strength or ductility of the wall.

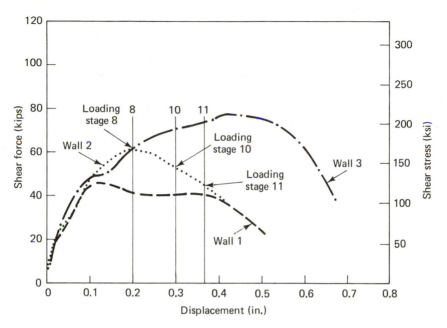

Figure 5.12 Force and stages versus displacement diagram [5.11]

Shear walls need to be designed for both shear and flexure. The distribution of the vertical (flexural) steel depends on the height-to-length (h/d) ratio. The presence of pilasters at the ends of the wall may also influence the location of the vertical steel. Squat shear walls ($h/d < 2$) are typically modeled as deep beams, while tall shear walls ($h/d > 2$) are more appropriately modeled as cantilever beams. The difference between the two models is best illustrated by the suggested distribution of wall steel.

Shear walls in low-rise structures usually have h/d ratios of less than 2 and are most commonly represented by a deep beam because of the geometric similarities between the two. As discussed in Section 3.6, the application of vertical load to the top edge and the application of the reactions directly to the bottom edge of a simply supported deep beam enhance the effectiveness of the arch action mechanism in carrying the applied load. This type of action renders ineffective the shear stirrups crossing the diagonal cracks because no compression struts can form between stirrup anchorages. These tension and compression struts are necessary parts of traditional shear-carrying mechanisms. It is shown that increasing the number of stirrups did not improve shear strength because the deep beam load is carried by the stiffest of the two possible mechanisms, and arch action is stiffer than truss action.

In both high- and low-rise shear wall buildings, the shear load is transferred from the diaphragm to the shear wall, at the floor-to-wall connection, as a line load. Figure 5.13 shows a typical low-rise shear wall and two free bodies located at different points. In the corner of the shear wall, Figure 5.13 illustrates the forces developed in the reinforcement and shows that both horizontal and vertical shear steel are required to resist the applied load. On the other hand, the resolution of the applied shear force at the center of the shear wall requires only that vertical forces, equal to the applied shear, be developed to generate the diagonal compression strut. The vertical steel can develop this vertical force, and, although commonly considered *shear steel*, is in reality functioning as flexural reinforcing resisting the overturning moment on the wall.

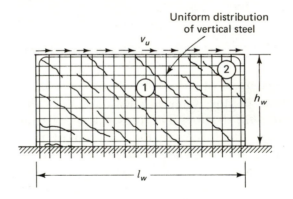

(a) Wall Elevation

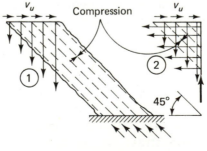

(b) Detail

Figure 5.13 Shear resistance of low-rise shear walls [5.10]

It is clear, then, that the vertical steel in a low-rise shear wall should be distributed in a fairly uniform pattern across the length of the wall, with only a slight increase at the vertical edges. This distribution is not applicable for tall shear walls, which tend to behave like cantilever beams. Since a ductile failure mechanism is desirable in shear walls, the flexural capacity should not exceed the shear capacity of the wall, and particularly in high-rise shear walls, the load required to produce a flexural failure should be one that causes relatively low shear stresses in the masonry. The increase in shear strength usually observed with increased axial load will most probably not occur in low-rise shear walls because of small gravity loads and the potential for vertical accelerations. Consequently, walls should be designed to resist the applied shear and no reliance should be placed on axial compression in the masonry.

The uniform distribution of vertical (flexural) steel suggested for a squat shear wall is not an efficient arrangement in a tall wall. Such an arrangement does not efficiently utilize the steel across the entire depth of the member. In shear walls subjected to large bending moments, the most efficient section will generally result from placing the flexural steel at the two ends because this increases the internal moment arm. It has been suggested that the wall steel required by the design moment be distributed such that minimum wall steel is placed over 80 percent of the length and the remainder of the required steel is placed over the remaining 10 percent region at each edge. The steel located at the end of the wall is referred to as *chord steel*.

The moment capacity of a shear wall is assumed to be that of a singly reinforced beam. In other words, the compression steel present at the end opposite to the tension steel is ignored unless it is placed in a pilaster and tied so as to prevent lateral buckling of the reinforcing steel. Thus, the flexural capacity of the wall shown in Figure 5.14 is

$$\phi M_n = \phi A_s f_y \left(d - \frac{a}{2}\right) \geqslant M_u \qquad (5.48)$$

with the design moment being equal to M_u.

In summary, the procedures for designing shear walls of concrete masonry should be consistent with the following

1. Factored loads should be used to determine the required strength, V_u.

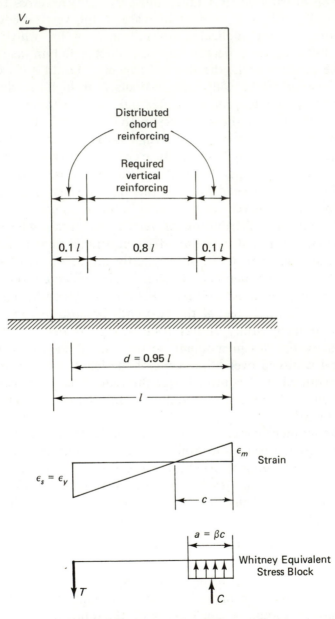

Figure 5.14 Flexure in tall shear walls

2. The nominal shear strength of reinforcing for walls should be a function of the aspect ratio (h/d) of the wall (see Section 3.5).

3. For short walls, $h/d < 1$, horizontal and vertical reinforcing should be approximately the same with little, if any, additional boundary flexural reinforcing provided. Shear carried by the concrete masonry increases significantly as the h/d ratio decreases.

4. For tall walls the flexural capacity of the section should exceed the shear strength. Horizontal reinforcing is the principal shear transfer material because truss action is expected to carry the shear load.

6

FUNDAMENTAL CONSIDERATIONS
IN DESIGN CRITERIA DEVELOPMENT

6.1 GENERAL

An essential concept in structural design is to produce a structural
system with sufficient capacity to resist the effects of the anticipated
loads imposed on it during the life of the structure. Although this
is a very straightforward concept, problems occur when one attempts
to establish the magnitude of the "anticipated loads" for which the
structure must be provided with "sufficient capacity" to resist. It
appears, then, that there are two fundamental problems which must
be resolved so that safe buildings may be constructed at economical
costs: What are the anticipated loads, and how should the capacity
of a structural member be established?

There is uncertainty associated with most aspects of the structural design and construction process. For example, structural
engineers cannot establish with certainty the maximum loads to
which a structure will be subjected during its life. Reinforcing steel
is not placed exactly as shown on the construction documents.
Structural engineers must design buildings in the face of this uncertainty and must do so with a final level of risk that is acceptable to
society.

The incorporation of uncertainty in structural problems suggests the use of probability theory. It is clear, however, that it is
not reasonable within today's level of practice to expect the designer
to attempt to quantify the level of each source of uncertainty.
Therefore, the development in the last decade of the *probability-
based limit state design* (PBLSD) approach has been intended to be
the rational vehicle for satisfying the requirements of practicality
and the probabilistic aspects of the design process.

The concept of the PBLSD method differs from the working
stress design (WSD) or deterministic strength design methods currently in use. The WSD method limits the effect of the specified
loads to a prescribed fraction of the true yield stress, buckling load,
compressive strength, or tensile strength of a particular material.
The factor of safety for a given stress state is the reciprocal of the
allowable fraction of the *material strength*. This factor of safety
has been subjectively determined from an examination of a wide
variety of structures that have been observed to perform satisfac-
torily throughout their useful life. The deterministic strength design
method requires that the effect of factored loads (i.e., actual service
loads modified by a load factor to produce a critical loading on the
member) not exceed the factored strength of the member. The
implicit incorporation of safety and reliability considerations in this
method is accomplished through the selection of the factors that
modify the load and strength effects and these factors are established
in a subjective manner.

The PBLSD method uses the design equation

$$(\text{Design resistance}) \geq (\text{effect of design loads}) \qquad (6.1)$$

However, in the development of the right- and left-hand sides of
Equation (6.1), uncertainty and probabilistic concepts are explicitly
incorporated into the design process. Limit states must be identified
by the structural engineer as a part of the process of designing the
structural system. The design format of the PBLSD method takes
the form [6.1]

$$\phi R_n > \sum_{i=1}^{n} \gamma_i Q_i \qquad (6.2)$$

where ϕ = strength reduction factor

R_n = calculated nominal capacity computed according to a prescribed formulation in the material specification using specified material strength and dimensions

γ_i = load factor

Q_i = service loads

The right-hand side of Equation (6.2) represents the summation of factored service loads specified by the appropriate building code for dead, live, wind, seismic, and other relevant loads. The specified load factors are intended to account for unfavorable variations inherent in the randomness and uncertainty associated with the true loads on a structure. It is especially important to note that *they are the same for all construction materials and ultimate limit states.* The load factors are derived from a probabilistic study of the distribution of these serivce loads in such a way as to maintain consistent levels of safety. One would expect, for example, that the load factor associated with live loads would be greater than that associated with dead loads because of the greater uncertainty and variability in establishing the magnitude of the live loads that may occur during the life of a structure.

The left-hand side of Equation (6.2) represents the computed nominal capacity of a limit state multiplied by a strength reduction factor. The nominal capacity of a limit state of a concrete masonry component is established using the principles presented in Chapters 3 and 5. The value of the strength reduction factor is a function of many items, including the building material, the limit state under consideration, the consequence of a particular type of failure, and the possible modeling errors.

Limit state refers to a situation in which a structural element or a structural system no longer satisfies its intended design objective. Designers typically consider two types of limit states: ultimate limit states and serviceability limit states. A limit state related to structural collapse is an *ultimate limit state.* This type of limit state is also called a strength or safety-related limit state. A *serviceability limit state* relates to functional utility. For example, even though a structural member may be structurally adequate from a collapse viewpoint, a serviceability limit state may exist because its utility has been severely reduced.

Limit state design requires that the designer explicitly consider possible limit states in a member or the limit states in an entire system. For example, such consideration might involve determining the load at which a wall under combined loading would fail in flexure and ascertaining that it is less than the load that will produce an undesirable compressive failure in the masonry. Most material specifications that employ the strength design method are formulated to produce the more ductile modes of failure, but it remains for the designer to step back and consider how a particular system might fail during an overload condition or because of an under-strength condition in one construction material.

The specification of both the load and strength reduction factors depends on the level of structural reliability deemed sufficient by those responsible for specifying these factors. This determination indirectly represents a determination by society as a whole as to what is adequate structural performance. The estimate of reliability is defined by the *reliability index, β*. New material specifications based on the PBLSD method may be introduced to provide reliability levels similar to those in the present codes by calibrating the new specification to the existing code. This might be accomplished by repeatedly establishing the reliability index for a wide variety of common design situations where past experience has demonstrated satisfactory performance. From this investigation a set of target reliability indices for a PBLSD criterion may be developed from which appropriate load and strength reduction factors are determined.

The remainder of this chapter discusses the concepts of structural reliability, load factors, and strength reduction factors in the context of PBLSD concepts.

6.2 STRUCTURAL RELIABILITY

The traditional approach to safety in structural design has essentially relied on judgment and common sense to derive the basic provisions in contemporary building codes and material specifications. Design was seen as an essentially deterministic process in that the uncertainty known to exist in the real world was not explicitly incorporated into the development of the code-specified design methods. Section 6.1 suggests that deterministic approaches in structural engineering are not entirely rational because the nature of structural resistance and loading is probabilistic. It then follows that the application of probabilistic concepts to structural problems should result in a model which more closely approaches the true behavior present in the real world.

It is also suggested in Section 6.1 that it is virtually impossible to avoid the possibility of structural failure, whether that failure be described as a strength or a serviceability failure. Probability theory can provide an estimate as to the likelihood of failure from which judgments may be made as to whether or not this represents an acceptable level of risk. The application of probability theory can accomplish this by developing a framework in which the available facts are used and decisions related to structural safety are made in a logical and clearly understood manner.

If one considers the shear wall in Figure 6.1, it may be seen that the nominal design strength of this member may be calculated using the appropriate limit state equations from Chapters 3 and 5. This design strength is considered to be a deterministic quantity in that the equation gives the designer but one value of moment capacity for a given set of dimensions, area of reinforcement steel, and material strengths. This number, however, does not represent the value of the *actual* moment strength, and until the member is loaded to failure, only statements of a probabilistic nature can be made regarding the beam's true strength. Similarly, only probabilistic statements

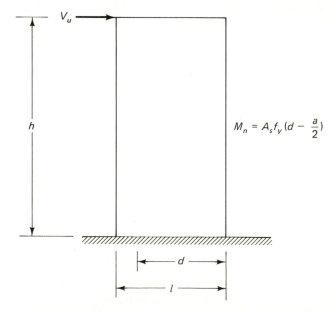

Figure 6.1 Nominal moment capacity of a shear wall

may be made regarding the actual loads that might be imposed on the shear wall during its design life.

The structural engineer must, however, make predictions regarding the anticipated loads which the member will experience as well as attempting to establish the capacity of the member for a given limit state. Probability is useful in making these predictions, in that probabilistic methods explicitly recognize that all predictions of the future have some level of uncertainty associated with them. These methods model reality by recognizing the observed scatter, randomness, and uncertainty present in actual designs, and quantify it using probability theory.

The distribution of the resistance, R (moment capacity, for example), offered by the wall in Figure 6.1 may be represented by using the *probability density function* (PDF) shown in Figure 6.2a, where the average or mean value of the resistance is given by $\overline{R}$.

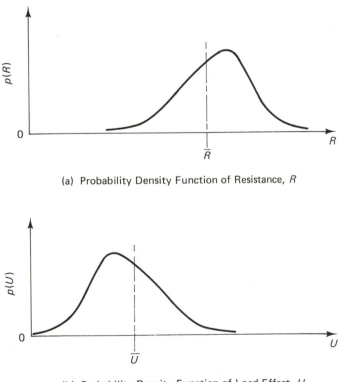

(a) Probability Density Function of Resistance, R

(b) Probability Density Function of Load Effect, U

Figure 6.2 Probability density functions of r and u

The horizontal axis represents the values of R. The vertical axis provides the ordinates to calculate the probability of a value of R having a value between R_i and R_j. The probability of R falling between R_i and R_j is equal to the area under the PDF between R_i and R_j. In a similar manner, the distribution of the load effects, U (moment induced by applied loads, for example), may be represented as shown in Figure 6.2b. The mean value of the load effect is given by $\overline{U}$. Thus, for a given range of U on the horizontal axis, one may find the associated probability $[P(u)]$ that such a range of U will occur. A more complete treatment of probability density functions is presented in reference [6.2].

If failure is described by the condition where the capacity of the member, R, is equal to or exceeded by the specified load effect, U, then failure occurs when R minus U is less than or equal to zero; that is,

$$F = R - U \qquad\qquad (6.3)$$

where F is the safety margin.

Thus, failure occurs when $F \leqslant 0$. This situation may be represented graphically by considering the two PDFs shown in Figure 6.2. If the two are superimposed on the same set of axes, as shown in Figure 6.3a, the shaded area represents the condition where $(R - U) \leqslant 0$. If the two PDFs representing $R - U$ in Figure 6.3a are then expressed as one PDF, as in Figure 6.3b, the probability of failure is the shaded area to the left of zero [6.2].

It may be seen from Figure 6.3a that the failure condition requires two separate events to occur before the member is judged to have failed because failure is a function of both capacity *and* load. Consequently, failure occurs when a member of moderately low strength is loaded with a very high level of load or a very low strength member is loaded with a moderately high load. As a result, it may be seen that the occurrence of an extremely high load does not necessarily represent a failure condition unless combined with a member of sufficiently low capacity.

The variability of the data about the mean value of the safety margin, $\overline{F}$, is quantified by the standard deviation, σ_F. The standard deviation represents a measure of the spread of the data. A given value of F may be described by how many standard deviations it is away from the mean. Thus, the mean of F is zero standard devia-

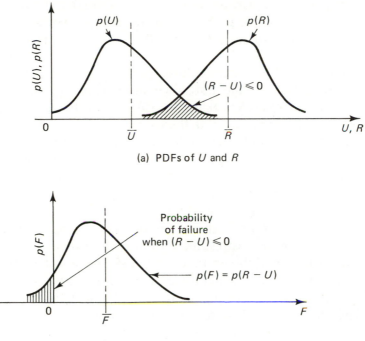

(a) PDFs of U and R

(b) PDF of the Safety Margin, F

Figure 6.3 Probability of failure

tions from the mean, while an extreme value of F might be three or four standard deviations (above or below) the mean. It is assumed that for a given value of F, the greater the number of standard deviations it is above or below the mean, the lower the probability that such a value of F will occur. The more unlikely it is that a value of F will be less than or equal to zero, the more unlikely it is that the member under consideration will fail. It thus possesses greater reliability.

If the values of $\overline{F}$ and σ_F are known, it is possible to define another term which gives an indication of the reliability of a particular element or structural system. The *reliability index*, β, is defined as

$$\beta = \frac{\overline{F}}{\sigma_F} \tag{6.4}$$

The reliability index has two fundamental advantages over conventional methods of reliability analysis. *It allows the strength of build-*

ing components to be viewed on a material by material basis, and it provides for and encourages the characterization of strength to be done independently of load factors. In addition, it enables one to address safety and reliability without directly quantifying the probability of component or system failure.

The advantage of the last observation may be more clearly understood if one considers that the load and resistance effects leading to structural failure occur at the extreme ends the PDFs describing R and U. The probability of failure is very sensitive to the PDF used to describe the distribution of the values of resistance and load effects because of the influence of the values at the extremes. The selection of different PDFs may result in changes in the probability of failure by several orders of magnitude. By avoiding the explicit specification of the probability of failure and relying on the reliability index, a more robust estimate of structural reliability may be obtained. It has been shown that most designs are not particularly sensitive to the actual probability of failure, and that measures of reliability not heavily dependent on the extreme tails of the PDFs describing the structural system should be used [6.3]. The reliability index is such a measure.

The reliability index is a measure of structural reliability, and with only the mean and standard deviation of the safety margin, the value of β may be determined. The reliability index indicates how many standard deviations below the mean of F are required before combinations of load and resistance effects will lead to structural failure, as shown in Figure 6.4. In other words, the greater the value of β, the greater the structural reliability, and the smaller the probability of failure. Typical values of β present in current masonry, concrete, and steel design codes, are shown in Table 6.1. It may be

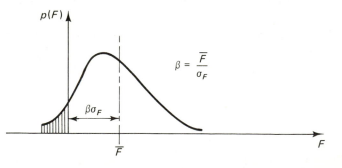

Figure 6.4 Reliability index β

TABLE 6.1   Typical Values of the Reliability Index, β,
from Current Design Codes [6.3]

	Masonry	Reinforced Concrete	Steel (Ultimate)
Beams	7.5–8.5	2.6–3.8	2.7–4.6
Columns	6.0–7.5	2.6–4.3	1.9–3.0

seen that the reliability index is much larger in masonry components than it is in the equivalent steel or concrete element. Additional research is required to assess the impact of strength design concepts on masonry structures, but it appears at this time that these values of β are too conservative. For example, it has been found that a value of $\beta = 3.0$ is consistent with average current practice for load combinations involving dead and live or dead plus snow loads, while $\beta = 2.5$ and 1.75 were representative for combinations describing wind and seismic loads, respectively. Therefore, current masonry β values are too large.

It is pointed out that the safety margin is dependent on the values of the load and resistance effects, and it is possible to express the value of the β index in terms of R and U if R and U are uncorrelated random variables. If the mean and standard deviation of the resistance are given by $\overline{R}$ and σ_R and the mean and standard deviation of the load effects are given by $\overline{U}$ and σ_U, the values of $\overline{F}$ and σ_F are

$$\overline{F} = \overline{R} - \overline{U} \qquad (6.5a)$$

and

$$\sigma_F = \sqrt{\sigma_R^2 + \sigma_U^2} \qquad (6.5b)$$

Thus, it can be seen that substituting Equations (6.5) into Equation (6.4), the expression for β may now be expressed as

$$\beta = \frac{\overline{F}}{\sigma_F} = \frac{\overline{R} - \overline{U}}{\sqrt{\sigma_R^2 + \sigma_U^2}} \qquad (6.6)$$

It should be noted that the variability of the safety margin is greater than that of either of the two components, R and U. This should be expected because F itself is a random variable and its variability depends directly on the spread of the data describing both R and U.

The reliability index can be used to establish the load and resistance factors used in the probability-based limit state design method introduced in Section 6.1. In the remaining two sections in Chapter 6 we describe the philosophy of the load and resistance factors and how they might be established for use in design.

6.3 LOAD FACTORS

Current material specifications use different load factors for different materials in attempting to describe the design load for the right-hand side of Equation (6.2). The most familiar load factors are those applied to the design of concrete structures as specified in the ACI Code [6.4]. Part II of the AISC Specification [6.5] lists the recommended load factors for use in the design of ductile steel frames. Materials designed with a working stress approach might be said to employ load factors of unity in assessing the magnitude of the design load. It would seem, however, that a single set of load factors could be developed for all materials in a manner that would simplify the structural design of these construction materials and promote a more uniform reliability than exists in current material specifications. In addition, relevant loading combinations could be specified to indicate those combinations that are most likely to occur in actual service conditions.

There are several problems that make the development of a unified set of load factors a difficult task. Current design criteria for the different materials result in different values of β. An important parameter in the variation of β is the ratio of live load to dead load, L/D. Within a particular material specification, different magnitudes of L/D also result in varying values of the reliability index. Other problems are related to the manner in which the description of the live load is simplified for use in the design of a structural system. The random (probabalistic) nature of all loads, in particular live, seismic, and wind loads, creates additional problems in deriving a single set of load factors and loading combinations. The National Bureau of Standards (NBS) has recently completed a study of this problem and has proposed a new set of load factors and loading combinations which have been adopted as part of ANSI A58.1 [6.6].

It may be seen in Figure 6.5 that the value of β varies with different L/D ratios for different materials. As the ratio of live load to dead load increases, the general trend is a reduction in the relia-

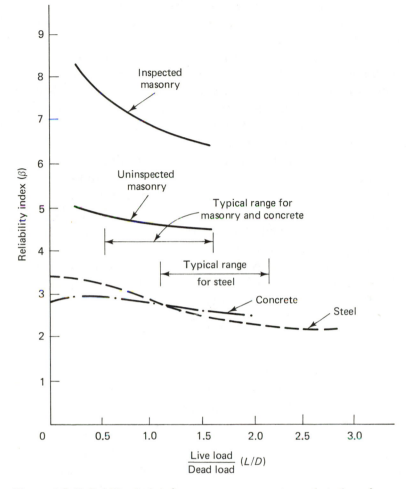

Figure 6.5 Reliability index for masonry, concrete, and steel conforming to current criteria

bility of the member. This is an important point to remember because as the live load becomes more dominant, the system is now being significantly influenced by a type of load with a large degree of variation. Further, the range of live load-to-dead load ratios common to most masonry and concrete structures is different from that found in typical steel structures. This occurs because the dead load is usually a larger portion of the design load in masonry and concrete buildings than in steel buildings. Any unified set of load factors must reconcile these differences between materials to achieve a more uniform level of reliability because different values of the load factor would be chosen if the probabalistic methodology were applied to each material independently.

Most live loads are specified as a uniformly distributed load even though the load is known to be randomly distributed with respect to time, location, orientation, and magnitude. It is doubtful that design codes will alter this method of describing live loadings in the foreseeable future, and the derivation of the load factors must account for this modeling error.

The last major consideration is the determination of reasonable load combinations to be used in design. It is unlikely that the maximum value of the dead, live, wind, snow, and seismic loads will all occur simultaneously. It is not reasonable to apply "overall" load factors to combined loadings to account for this unlikely event, as is done in some codes. For example, if one were considering a seismic plus a dead- and live-load combination, the use of the following load-factor format [6.7] might result in a poor approximation of the effects of the combined loads:

$$U = \gamma_D D_n + \psi(\gamma_L L_n + \gamma_W W_n + \gamma_T T_n) \qquad (6.7)$$

where γ = load factor

ψ = load combination probability factor equal to 1.0, 0.7, or 0.6, depending on whether one, two, or three loads are included in the brackets

D_n, L_n, W_n, T_n = loads

For the loading case in Equation (6.7), if one considers an interior column in a lateral-load-resisting frame, the vertical loads are from the combined dead and live load while the moment is primarily due to the lateral load. Thus, if the critical loading combination must consider both axial load and moment in the columns from $D + L + W$, the designer would end up designing for only 70 percent of the applied lateral-load moment. The simpler approach adopted in Equation (6.2) appears to represent the actual effects of combined loads better than does Equation (6.7).

Figure 6.6 illustrates the combination of different types of loads in a structure. If the load is broken up into permanent (dead) loads, sustained (live) loads, and loads of short duration (seismic, wind), the maximum total load will probably occur at some point that does not represent the combination of the maximum values of

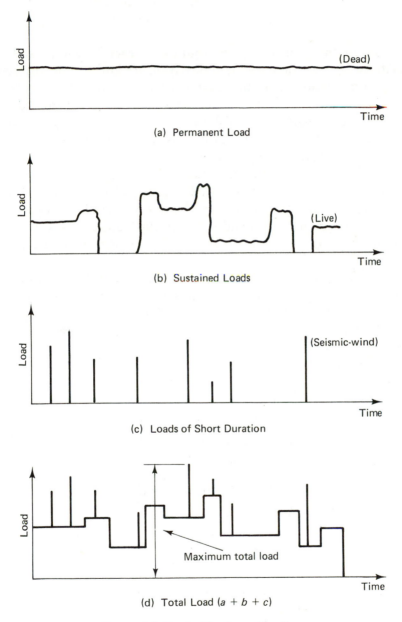

(a) Permanent Load

(b) Sustained Loads

(c) Loads of Short Duration

(d) Total Load $(a + b + c)$

Figure 6.6 Typical load combinations

the individual loads. The time interval over which the load is monitored for the *maximum* combination is typically assumed to be the building design life, which, in the NBS study, was taken to be 50 years.

Now, if the probability distribution of each load type and an estimate of its mean and standard deviation are known, the different loading combinations may be simulated and the strength of various members computed. If the calculated strength is compared to the actual tested strength, the reliability of the section may be computed. The probability distributions used in the derivation of the NBS load factors are shown in Table 6.2.

It is clear that for each material, load combination, and structural component (i.e., column, wall, beam), different values of β will be found. The basic idea is to identify a target value of β for a set of loading combinations and establish the load factors which will allow the value of β to remain as constant as possible under those conditions.

Assume that it is desired that β be equal to 3.0 for load combinations involving dead plus live load or dead plus snow load for all materials and structural components and β be equal to 2.5 and 1.75 for load combinations involving wind and seismic loads, respectively. Then it follows that the load combinations and load factors can be derived, and those obtained in the NBS study [6.3] are

$$U = 1.4D \tag{6.8a}$$

$$= 1.2D + 1.6L \tag{6.8b}$$

$$= 1.2D + 1.6S + (0.5L \text{ or } 0.8W) \tag{6.8c}$$

$$= 1.2D + 1.3W + 0.5L \tag{6.8d}$$

$$= 1.2D + 1.5E + (0.5L \text{ or } 0.2S) \tag{6.8e}$$

$$= 0.9D - (1.3W \text{ or } 1.5E) \tag{6.8f}$$

TABLE 6.2   Load Parameters [6.3]

Load	PDF
Dead	Normal
Live	Type I extreme value
Wind	Type I extreme value
Snow	Type II extreme value
Earthquake	Type II extreme value

It may be seen from Equations (6.8b) through (6.8e) that the factored load combinations modify one variable load at a time while maintaining a constant load factor on the dead load. The load factor of 1.4 in Equation (6.8a) is to provide a minimum reliability when the dead load completely dominates all the other load combinations. Equation (6.8f) might govern in conditions where there is significant axial tension because of overturning forces from earthquake or wind loads. In these situations it might be nonconservative to overestimate the dead load. The dead load may not be as great as predicted because of the conservative assumptions often made in estimating the dead load as well as the apparent reduction in downward force caused by vertical accelerations during earthquakes.

It is interesting to note that the load factor for the dead load derived from the regression analysis was actually on the order of 1.1 rather than the 1.2 specified in Equation (6.8). However, it was felt that a value of 1.1 was too low to be accepted by the structural engineering profession.

The load combination and load factors in Equations (6.8) represent an attempt to incorporate probabalistic concepts into the derivation of these load factors, as well as an attempt to achieve a more uniform level of structural reliability than exists in current material specifications and design codes. These load factors have been adopted and are now part of the American National Standards Institute ANSI A58.1 (1982).

6.4 STRENGTH REDUCTION FACTORS

The strength reduction factor, ϕ, on the left-hand side of Equation (6.2) is an attempt to account for the variations in the in-situ capacity of the member from that calculated using the analysis and design equations presented in Chapters 3 and 5. If the designer had the ability to establish the exact capacity of a structural section, it would be straightforward to compare the calculated resistance with the anticipated maximum load effects. If the capacity were greater than the demand, the section would be considered adequate. Unfortunately, just as a structural engineer does not have the ability to establish with certainty the maximum design loads, neither is it possible to determine the exact capacity of a section for each limit state under consideration.

We have discussed previously the fact that much of what is currently considered to be deterministic in design is actually a random quantity. The structural engineer may specify an ultimate compressive strength, f'_m, of 2.0 ksi for the masonry. Lab tests of f'_m will probably not equal 2.0 ksi but instead will vary in a random manner about some mean value of the compressive strength, $\bar{f}'_m$. In addition, the true in-situ compressive strength of the masonry will very likely match neither the specified value of f'_m nor the lab-tested values of f'_m. In a similar manner, the actual yield strength of the steel will vary from that specified, and the actual and specified dimensions will not be the same for most members in the structure. Most analytical and design models incorporate many assumptions and simplifications intended to make them more convenient for the designer to use. Whether or not these assumptions are of a conservative or nonconservative nature is generally not of much significance in establishing the actual capacity of a structural member, inasmuch as the assumptions simply introduce another source of error. Of course, from the standpoint of the designer, conservative assumptions represent one method of acquiring confidence in one's design in the face of an acknowledged short supply of good, accurate information.

The fact that all of the strength reduction factors are less than unity implies that all the variations in material strengths, workmanship, and modeling considerations are of an undesirable nature. This is not true, but prevailing philosophy in structural design is to try to err on the side of conservatism.

The ϕ factor may also be used to reflect the consequence of failure for a given structural member. It is generally conceded that the failure of a column, whether it be a ductile failure or not, represents a very undesirable condition. This is so because columns are typically used to support several beams and possibly column loads from the floors above. The column may also be part of a lateral force resisting element such as a frame. The failure of a column generally leads to the failure of the structural elements to which it is attached and may be the direct cause of a system-wide structural collapse. The failure of an isolated beam will typically lead to failure only in the immediate influence area of the beam, and usually is not cause for a system-wide collapse.

Thus, it seems reasonable to assume that the structural reliability of columns failing in compression should be higher (i.e., a

TABLE 6.3 Strength Reduction Factors for Concrete [6.4]

Action	Factor
Flexure, with or without axial tension	0.90
Axial tension	0.90
Axial compression, with or without flexure	
Members with spiral reinforcement	0.75[a]
Other reinforced members	0.70[a]
Shear and torsion	0.85
Bearing on concrete	0.70
Flexure in plain concrete	0.65

[a]May be increased linearly to 0.90 as ϕP_n decreases from $0.10 f'_c A_g$ or ϕP_b, whichever is smaller, to zero.

smaller ϕ factor) than that associated with elements where a flexural failure is anticipated. The ϕ factors recommended in the current ACI Code bear this point out. It may be seen in Table 6.3, for example, that the strength reduction factor for beams is greater than that for columns failing in compression. This same philosophy permits the linear increase in the ϕ factor from 0.7 to 0.9 when the failure of the column is controlled by flexure under light axial loads.

Recall from Equation (6.4) that the value of β may be found by determining the mean and standard deviation of the safety margin, F:

$$F = R - U \qquad (6.9)$$

and

$$\beta = \frac{\overline{F}}{\sigma_F} = \frac{\overline{R} - \overline{U}}{\sqrt{\sigma_R^2 + \sigma_U^2}} \qquad (6.10)$$

Neither the resistance nor the applied load effects may be determined without the introduction of uncertainty. If it is assumed that the design load demand is modified by a probabilistically defined coefficient, then the actual load demand, U, could be written

$$U = X_1 U_D \qquad (6.11)$$

where U = actual load demand

U_D = design load demand (a known constant)

X_1 = a variable, with uncertainty, that is used to scale the design load demand to obtain the actual load demand

The random variable X_1 can therefore be expressed in terms of other random variables to incorporate various factors which may account for the variation between the design load and the actual load. These other random variables, or *base random variables*, may be used to describe a function that will provide the value of X_1. Thus, X_1 might be expressed in terms of the following base random variables:

$$X_1 = f(Y_1, Y_2, \ldots, Y_n) \tag{6.12}$$

where $Y_{1,n}$ are the random variables incorporating the uncertainty affecting the load demand. The base random variables might consider the uncertainty in the dead load, live load, structural engineering model, and the like.

The uncertainty associated with the capacity of the member may be expressed in a manner similar to that used to account for the uncertainty in the load demand:

$$R = X_2 R_D = X_2 \phi R_n \tag{6.13}$$

where R = actual member capacity

   R_D = calculated design capacity

   R_n = calculated nominal member capacity

   X_2 = a variable, with uncertainty, that is used to scale the calculated design capacity to obtain the actual member capacity

   ϕ = strength reduction factor

Of course, the random variable X_2 can be expressed in terms of other base random variables which incorporate the uncertainty in the variability between the actual and calculated member capacity. Thus, X_2 might be expressed in terms of the following base random variables:

$$X_2 = f(Z_1, Z_2, \ldots, Z_n) \tag{6.14}$$

where $Z_{1,n}$ is the random variable incorporating the uncertainty affecting the member capacity. The base random variables might consider the uncertainty in the laboratory tested capacity compared to the calculated member capacity (R/R_n), workmanship, material variations, professional error, and the like.

Recalling again the definition of the safety margin, failure will be defined when F is less than or equal to zero. If equations (6.11) and (6.13) are combined, the safety margin may be written as

$$F = X_2 \phi R_n - X_1 U_D \tag{6.15}$$

and the mean value of the safety margin is then

$$\overline{F} = \overline{X}_2 \phi R_n - \overline{X}_1 U_D \tag{6.16}$$

where $\overline{F}$ = mean value of the safety margin

$\overline{X}_{1,2}$ = mean value of X_1 and X_2, respectively

It can also be shown that the standard deviation of F is

$$\sigma_F = \sqrt{(\phi R_n)^2 \sigma_{X_2}^2 + U_D^2 \sigma_{X_1}^2} \tag{6.17}$$

where σ_F = standard deviation of F

$\sigma_{X_1}, \sigma_{X_2}$ = standard deviation of X_1 and X_2, respectively

Referring to Equation (6.10), the reliability index, β, may be expressed as

$$\beta = \frac{\overline{X}_2 \phi R_n - \overline{X}_1 U_D}{\sqrt{(\phi R_n)^2 \sigma_{X_2}^2 + U_D^2 \sigma_{X_1}^2}} \tag{6.18}$$

If probability distribution functions, means, and standard deviations of Y_i and Z_i, *or alternately X_1 and X_2*, are established, it is then possible to establish the value of ϕ given the value of β. Note that uncertainties in loading and capacity are present in Equation (6.18).

The topic of the development of ϕ factors is one that has received considerable attention and certainly merits further consideration from both the research and design communities. Appendix A presents a method of calibrating the strength reduction factors of one material to those of another material to achieve equivalent levels of structural reliability.

7

DESIGN CASE STUDIES

7.1 GENERAL

This chapter presents two case studies to illustrate the application of the theoretical development presented in the preceding chapters. The first example is the design of the masonry components for a one-story industrial-type building common to many parts of the country. A four-story shear wall is analyzed and designed in the second case study.

The intent of this chapter is to demonstrate the application of limit state design concepts as they might be applied to masonry structures. The design of reinforced concrete block masonry is currently performed using the working stress design (WSD) method, and the actual and philosophical differences between limit state design and the WSD method are presented and discussed in the preceding chapters.

In light of the scarcity of experimental data for some aspects of limit state design as applied to masonry structures, some general assumptions have been made in the examples in this chapter. Perhaps the greatest assumption concerns the use of the strength reduction factors (ϕ factors). The development of strength reduction factors is discussed in Chapter 6. For a given limit state and reliability index, a value of ϕ may be chosen using the concepts presented in that chapter. For the purpose of illustration, however, the mean ϕ factors derived in Appendix A will be used. Although the actual values of ϕ may change when limit state design in masonry is accorded formal recognition in the building codes, the concepts and philosophy presented in the design case studies should not change appreciably.

Chapter 23 of the *Uniform Building Code* (UBC) is used to develop the design loads, both gravity and seismic. Not all jurisdictions use the UBC, and the use of local codes may result in different numerical values to be applied to the problems at hand. However, the differences are not of any substantial nature, inasmuch as the underlying fundamentals remain unchanged. The ANSI load factors discussed in Chapters 5 and 6 are used in the case studies to develop all factored loadings except for the factored shear wall load. To attempt to ensure that the flexural capacity of the wall is less than the shear capacity, the ACI load factor of 2.0 is used for shear walls.

The masonry and steel shear strengths used in the case studies are obtained from Table 5.3 in Volume 1. Since the publication of Volume 1, the International Conference of Building Officials have approved the use of strength design for shear walls using slightly lower values for V_m and V_n [7.1] in a research report sponsored by the Concrete Masonry Association of California and Nevada [7.2]. These changes will have some effect on the designs presented in the case studies; however, none of these modifications would result in a significant departure from the design method presented in this chapter.

7.2 ONE-STORY DESIGN CASE STUDY

Consider the design of the one-story concrete masonry building shown in Figures 7.1 and 7.2. The walls are assumed to be constructed of partially grouted 8-in. concrete block with an ultimate compressive strength of $f_m' = 1.5$ ksi. Grade 40 steel ($f_y = 40$ ksi) will be used in the design.

The roof dead load is assumed to equal 13 psf. The 1982 *Uniform Building Code* (UBC) is used to establish the roof live load. The value of the live load (LL) depends on the tributary area, A_T, according to the following schedule:

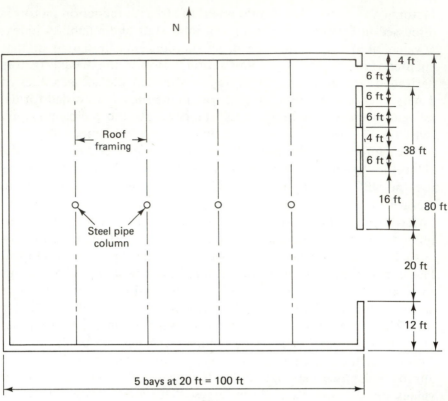

Figure 7.1 Building plan

$$A_T \leqslant 200 \text{ ft}^2 \qquad LL = 20 \text{ psf}$$

$$200 \text{ ft}^2 < A_T \leqslant 600 \text{ ft}^2 \qquad LL = 16 \text{ psf}$$

$$A_T > 600 \text{ ft}^2 \qquad LL = 12 \text{ psf}$$

The force levels for the seismic analysis are also determined from the 1982 UBC. It is assumed that the structure under consideration is located in seismic Zone 4, with a corresponding value of Z equal to 1.0. Masonry shear wall structures are classified as box-type structures, and the structural system characteristic factor, K, is equal to 1.33. The high value of K reflects the belief that these box structures possess little ductility, an assumption that has been shown not to be true. This value of K also accounts for the lack of redundancy present in

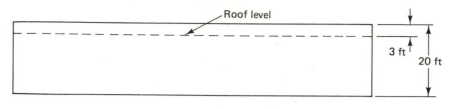

South Elevation

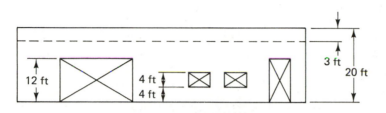

East Elevation

Figure 7.2 South and East elevations

structures where the walls resist both gravity loads and lateral loads. Inasmuch as there is no specific site-period information, the value of S is assumed to be equal to 1.5 and the limiting value of the product $CS = 0.14$ is used in the analysis.

Seismic:

$$\text{Zone 4: } Z = 1.0$$

$$\text{Commercial building: } I = 1.0$$

$$\text{Shear wall/box-type building: } K = 1.33$$

$$\text{Acceleration coefficient } C = \frac{1}{15\sqrt{T}}$$

80-Ft Direction:

$$T \simeq \frac{0.05(h)}{\sqrt{D}}$$

$$= \frac{0.05(20 \text{ ft})}{\sqrt{80 \text{ ft}}} = 0.11 \text{ sec}$$

$$C = \frac{1}{15\sqrt{T}}$$

$$= \frac{1}{15\sqrt{0.11}}$$

$$= 0.20 > 0.12 \qquad \text{Therefore, } C = 0.12$$

$$CS = 0.12(1.5)$$

$$= 0.18 > 0.14 \qquad \text{Therefore, } CS = 0.14$$

Building: $V = ZIKCSW$

$$= (1.0)(1.0)(1.33)(0.14)W = 0.186W$$

100-Ft Direction:

$$CS = 0.14$$

Building: $V = ZIKCSW = 0.186W$

For individual components, the value of the lateral load, applied perpendicular to the element, is computed using a different formula from that used to compute the lateral load for the structure as a whole. Note that the term CS is replaced by the value C_p, which varies depending on the structural element under consideration. For exterior walls the value of C_p is equal to 0.3.

$$V = ZIC_p W_p$$

$$= 1.0(1.0)0.30(W_p)$$

$$= 0.3W_p \text{ is used for out-of-plane loading}$$

East-West Wall Design for Out-of-Plane Bending

The load factors to be used are those discussed in Chapters 5 and 6. The loading model for flexure assumes a 4-in. eccentricity for the vertical gravity load, and a uniformly applied lateral load. The parapet is assumed to overhang from the point of support provided by the roof connection.

The design of the wall for out-of-plane bending is performed assuming that the wall is pinned at the top and bottom as shown in Figure 7.3. The height-to-thickness ratio (h/t) is checked to verify that the wall meets current limitations on h/t. It is assumed that an upper limit of $h/t = 36$ is permitted as discussed in Chapter 4. If

Wall Section

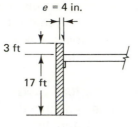

Out-of-Plane Bending

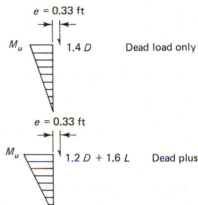

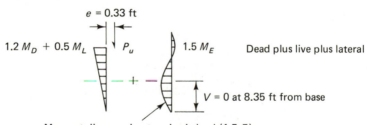

Moment diagram due to seismic load (1.5 E)

Figure 7.3 East and West bearing wall loading models

the wall did not meet this h/t ratio, an analytical model assuming a propped cantilever as described in Section 4.3 could have been tried. Additional steel would have to have been provided to carry the moment couple at the base of the wall.

$$\frac{h}{t} = \frac{17(12)}{8}$$

$$= 25.5 < 36 \qquad \text{Therefore, the } \frac{h}{t} \text{ is O.K.}$$

Using the loading combinations described in Section 5.1, the required ultimate wall flexural strength for out-of-plane loads is determined. The effect of axial loads is ignored when computing moments caused by application of the eccentric dead and live load and earthquake load.

Figure 7.3 shows the three load cases considered. An eccentricity of 4 in. for the roof load is assumed with a tributary width of 10 ft. Thus, the applied dead and live loads are equal to 130 plf and 160 plf, respectively. For the seismic load case, a uniform wall weight of 58 psf is applied up the height of the wall. The parapet is assumed to overhang from the point of support provided by the roof connection.

Vertical load (10 ft tributary width):

$$U = 1.2D + 1.6L$$

$$= 1.2(13 \text{ psf} \times 10 \text{ ft}) + 1.6(16 \text{ psf} \times 10 \text{ ft})$$

$$= 0.41 \text{ k/ft}$$

Lateral load (exterior bearing wall with load normal to flat surface; a wall self-weight of 58 psf for vertical steel at 24 in. on center is used): A comparison of lateral loads shows that seismic controls over wind.

$$U_W = 1.3W$$

$$= 1.3(15 \text{ psf})$$

$$= 0.020 \text{ ksf}$$

$$U_s = 1.5E$$

$$= 1.5(0.3)W_p$$

$$= 1.5(0.3)(58 \text{ psf})$$

$$= 0.026 \text{ ksf} \quad \text{governs}$$

Calculate the ultimate moment using load combinations.

Dead load only:

$$M_u = 1.4M_D$$

$$= 1.4(0.130 \text{ klf})(0.33 \text{ ft})$$

$$= 0.060 \text{ ft-kip/ft}$$

Dead and live loads:

$$M_u = 1.2M_D + 1.6M_L$$

$$= [1.2(0.130 \text{ klf}) + 1.6(0.160 \text{ klf})](0.33 \text{ ft})$$

$$= 0.136 \text{ ft-kip/ft}$$

Dead plus live loads plus earthquake (maximum moment from seismic load is assumed at 8.24 ft from base; effect of dead and live load on location of maximum moment is assumed to be negligible):

$$M_u = 1.2M_D + 0.5M_L + 1.5M_E$$

$$= [1.2(0.130 \text{ klf}) + 0.5(0.160 \text{ klf})](0.33 \text{ ft}) \left(\frac{8.24 \text{ ft}}{17 \text{ ft}}\right)$$

$$+ \frac{(0.026 \text{ ksf})(17 \text{ ft} + 3 \text{ ft})^2(17 \text{ ft} - 3 \text{ ft})^2}{8 \times (17 \text{ ft})^2}$$

$$= 0.92 \text{ ft-kips/ft} \quad \text{governs}$$

The design moment, M_u, equals 0.92 ft-kips/ft.

The out-of-plane moment capacity of the wall is checked by first assuming a rectangular beam analysis. It will be shown that the entire compression block, C, is in the face shell, which validates the assumption. The effective width of the wall, b, is taken to be the smaller of six times the block thickness (45.78 in.), or the center-to-center spacing of the vertical wall steel (24 in.). Vertical reinforcing is assumed to be No. 5 bars at 24 in. on center. Figure 7.4 shows the loaded wall system.

Equations (5.8) and (3.19) give the distance to the neutral axis.

$$a = \frac{A_s f_y}{0.85 f'_m b}$$

$$= \frac{0.31 \text{ in.}^2 (40 \text{ ksi})}{0.85(1.5 \text{ ksi})(24.0 \text{ in.})}$$

$$= 0.41 \text{ in.}$$

$$c = \frac{a}{0.85}$$

$$= \frac{0.41}{0.85}$$

$$= 0.48 \text{ in.} \leqslant 1.25 \text{ in.}$$

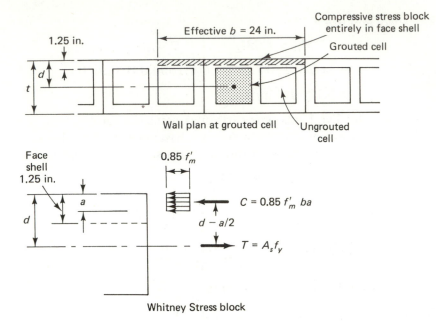

Figure 7.4 Model of wall for out-of-plane bending

Therefore, the compression block is in the face shell and a rectangular beam analysis is appropriate.

The strength reduction factor for flexure in a lightly reinforced beam is 0.80, as shown in Appendix A. Considering an effective width of 24 in. ($M_n = 0.92(2)12 = 22.08$ in.-kips) the reduced nominal moment, ϕM_n, is compared to the required ultimate moment, and shown to be adequate.

$$\phi M_n = \phi A_s f_y \left(d - \frac{a}{2} \right)$$

$$= 0.8(0.31 \text{ in.}^2)(40 \text{ ksi}) \left(\frac{7.63 \text{ in.}}{2} - \frac{0.41 \text{ in.}}{2} \right)$$

$$= 35.8 \text{ in.-kips} > 22.08 \text{ in.-kips}$$

A check is then performed to verify that the No. 5 vertical bars at 24 in. on center and a horizontal steel spacing of No. 5 bars at 48 in. on center meets the minimum steel requirements described in Section 2.5. (These steel amounts may have to be changed to carry the shear from the lateral loads.)

The steel ratios based on the gross area in the vertical and horizontal directions, ρ_v and ρ_h, respectively, are

$$\rho_v = \frac{0.31}{7.63 \text{ in.}(24 \text{ in.})} = 0.0017$$

$$\rho_h = \frac{0.31}{7.63 \text{ in.}(48 \text{ in.})} = 0.0008$$

The sum of the two steel ratios is shown to be greater than the required minimum of 0.002.

$$\rho_v + \rho_h = 0.0017 + 0.0008 = 0.0025 > 0.002$$

Lintel Design

The normal assumption, involving arch action in computing the vertical gravity load on the lintel, is ignored in the example (Figure 7.5). The long span, and the fact that the wall opening is relatively close to the top of the parapet, probably make such an assumption unjustified.

It will be noted that a check for both positive (moment at mid-span) and negative moment (moment at support) is made because the lintel is modeled as a fixed-fixed beam.

Roof dead load: 10 ft × 13 psf = 130 plf

Roof live load: 10 ft × 20 psf = 200 plf

Parapet and wall: 8 ft × 58 psf = 464 plf

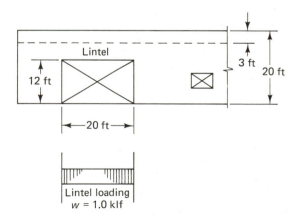

Figure 7.5 Partial East elevation at lintel

The factored uniform load and respective moments are calculated.

$$U = 1.2(130 + 464) + 1.6(200)$$

$$= 1.03 \text{ klf}$$

$$M_u^+ = \frac{wL^2}{24}$$

$$= \frac{1.03 \text{ klf}(20 \text{ ft})^2}{24}$$

$$= 17.2 \text{ ft-kips}$$

$$M_u^- = \frac{wL^2}{12}$$

$$= \frac{1.03 \text{ klf}(20 \text{ ft})^2}{12}$$

$$= 34.4 \text{ ft-kips}$$

Although the span to depth ratio probably makes a deep beam model of the lintel desirable, a slender beam approach will be used for simplicity. The reader should consider the implications of the differences between the two modeling assumptions. The distance available from the top of the wall to the opening is 96 in. The effective depth used is $d = 92$ in.

Equations (5.4) and (5.8) are used to calculate the moment capacity assuming No. 6 bar.

$$\rho = \frac{0.44 \text{ in.}^2}{7.63 \text{ in.} \times 92.0 \text{ in.}} = 0.001$$

$$a = \frac{A_s f_y}{0.85 f_m' b} = \frac{(0.44 \text{ in.}^2)(40 \text{ ksi})}{(0.85)(1.5 \text{ ksi})(7.63 \text{ in.})} = 1.81 \text{ in.}$$

$$\phi M_n = \phi A_s f_y \left(d - \frac{a}{2} \right)$$

$$= (0.8)(0.44 \text{ in.}^2)(40 \text{ ksi}) \left(92.0 \text{ in.} - \frac{1.81 \text{ in.}}{2} \right) \left(\frac{1 \text{ ft}}{12 \text{ in.}} \right)$$

$$= 106.9 \text{ ft-kips} > 34.4 \text{ ft-kips} \qquad \text{Therefore, O.K.}$$

The cracking moment of the section is computed to ensure that an amount of steel is provided to guarantee that the tension steel has a capacity at least as great as the tensile force generated by the cracking moment. This prevents the section from failing as the tension capacity of the section is reached. The tensile strength of the masonry is assumed to equal 0.1 ksi. The cracking moment is computed assuming an uncracked section; hence, the full value of the section modulus is used. Equation (3.14) is used to calculate the cracking moment.

$$M_{CR} = f_{mt}S$$

$$= \frac{(0.1 \text{ ksi})(7.63 \text{ in.})(92.0 \text{ in.})^2}{6}$$

$$= 89.7 \text{ ft-kips}$$

$$\phi M_n = 106.9 \text{ ft-kips} > M_{CR} = 89.7 \text{ ft-kips}$$

Therefore, additional steel is not required.

A check, using Equation (5.4a), is made to show that the steel ratio, ρ, is less than $0.75\rho_{bal}$. This check is to assure that the tension steel will yield before the masonry crushes. It is seen that the steel ratio is significantly below the maximum amount.

$$0.75\rho_{bal}$$

$$= \frac{(0.75)(0.85)f'_m \beta[0.002/(0.002 + f_y/E_s)]}{f_y}$$

$$= \frac{(0.75)(0.85)(1.5 \text{ ksi})(0.85)[0.002/(0.002 + 40 \text{ ksi}/29 \times 10^3 \text{ ksi})]}{40 \text{ ksi}}$$

$$= 0.012 > 0.001 \quad \text{Therefore, O.K.}$$

Verify that the depth of the lintel is sufficient to avoid shear reinforcing in addition to the minimum steel specified by the code. Refer to Table 5.3 of Volume 1.

$$f_{bearing} < 50 \text{ psi} \quad f'_m = 1500 \text{ psi}$$

$$V_u = \frac{wL}{2}$$

$$= \frac{(1.03 \text{ klf})(20 \text{ ft})}{2}$$

$$= 10.3 \text{ kips}$$

$$\frac{M}{Vd} = \frac{(34.4 \text{ ft-kips})(12 \text{ in./ft})}{(10.3 \text{ kips})(92.0 \text{ in.})}$$

$$= 0.44$$

$$\frac{4.0\sqrt{f_m'} - 1.7\sqrt{f_m'}}{0.25 - 1} = \frac{4.0\sqrt{f_m'} - V_m}{0.25 - 0.44}$$

Solving for the shear strength by linear interpolation gives $V_m = 0.132$ ksi and a comparison is made.

$$\phi V_n = \phi V_m \, db$$

$$= (0.68)(0.132 \text{ ksi})(92.0 \text{ in.})(7.68 \text{ in.})$$

$$= 63.4 \text{ kips} > V_u = 10.3 \text{ kips}$$

The masonry shear strength alone is greater than V_u and additional shear reinforcing is not required.

North/South Wall Design at Beams

Slenderness effects are considered in the design of vertical load-carrying walls. The effective width of the column (Figure 7.6) is assumed to be the width of the beam carrying the roof load plus four times the width of the wall. The entire effective width is grouted solid to develop an even load distribution.

A check is made to show that the applied vertical load is less than the axial capacity at the balance point using Equation (5.23). This implies that flexure controls the failure of the section, rather than compression.

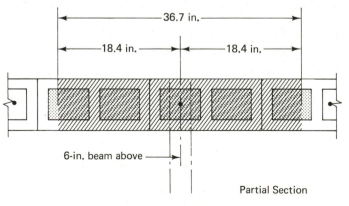

Figure 7.6 Wall-column

Roof beam load:

$$P_{u1} = 1.2(400 \text{ ft}^2 \times 13 \text{ psf}) + 1.6(400 \text{ ft}^2 \times 16 \text{ psf})$$

$$= 16.5 \text{ kips}$$

Wall load:

$$q = (36.7 \text{ in.})(7.63 \text{ in.})(0.15 \text{ kcf})/(1 \text{ ft}/12 \text{ in.})^2$$

$$= 0.29 \text{ klf}$$

$$P_{u2} = 1.2(0.29 \text{ klf})(10 \text{ ft})$$

$$= 3.5 \text{ kips}$$

Lateral wall load:

$$q_u = (1.5)(0.30q_p)$$

$$= (1.5)(0.30)(0.29 \text{ klf})$$

$$= 0.13 \text{ klf}$$

Try three No. 5 bars.

$$A_s = 0.93 \text{ in.}^2$$

$$a_b = \frac{58000\beta d}{(58,000 + f_y)}$$

$$= (58,000)(0.85) \left(\frac{7.63 \text{ in.}}{2}\right) \bigg/ (58,000 + 40,000)$$

$$= 1.92 \text{ in.}$$

$$P_b = 0.85 f_m' b a_b - A_s f_y$$

$$= 0.85(1.5 \text{ ksi})(36.7 \text{ in.})(1.92 \text{ in.}) - (0.93 \text{ in.}^2)(40 \text{ ksi})$$

$$= 52.6 \text{ kips}$$

$$P_{u1} \leqslant \phi P_b$$

$$\leqslant (0.65)(52.6 \text{ kips}) = 34.2 \text{ kips}$$

16.5 kips < 34.2 kips   Therefore, tension controls

The current ACI Code permits the ϕ factor to be increased from a compression-related value of ϕ to one that reflects the fact that flexure governs the design if the value of the axial load is less than the smaller of 10 percent of the maximum axial load ($0.1 f_c' A_g$) or

ϕP_b. If it is assumed that the same philosophy is reasonable in concrete masonry design, the revised value of ϕ may be determined from linear interpolation (the 3.2 factor in the interpolation equation represents the slope of the line between the two extreme values of the ϕ-factor). The appropriate values of ϕ are selected from Appendix A.

$$\frac{P_{u1}}{f'_m A_g} = 16.5 \text{ kips}/[(1.5 \text{ ksi})(36.7 \text{ in.})(7.63 \text{ in.})]$$

$$= 0.04 < 0.10 \quad \text{Therefore, reduction is O.K.}$$

$$\phi = 0.8 - \frac{3.2 P_n}{f'_m A_g}$$

$$= 0.8 - \frac{3.2(16.5 \text{ kips})}{(1.5 \text{ ksi})(281.6 \text{ in.}^2)}$$

$$= 0.68 > 0.64$$

The design moment strength is now calculated using the modified ϕ factor. The effective moment of inertia (I_e) is calculated to estimate the deflection due to the slenderness effects. The applied moment is calculated as shown in Section 5.3.

$$P_n = \frac{P_{u1}}{\phi}$$

$$= \frac{16.5 \text{ kips}}{0.68}$$

$$a = \frac{P_n + A_s f_y}{0.85 f'_m b}$$

$$= \frac{24.3 \text{ kips} + (0.93 \text{ in.}^2)(40 \text{ ksi})}{(0.85)(1.5 \text{ ksi})(36.7 \text{ in.})}$$

$$= 1.32 \text{ in.}$$

$$M_n = (P_n + A_s f_y) \left(d - \frac{a}{2}\right)$$

$$= [24.3 \text{ kips} + (0.93 \text{ in.}^2)(40 \text{ ksi})] \left(3.82 \text{ in.} - \frac{1.32 \text{ in.}}{2}\right) \left(\frac{1 \text{ ft.}}{12 \text{ in.}}\right)$$

$$= 16.3 \text{ ft-kips}$$

$\phi M_n = 0.68(16.3 \text{ ft-kips})$

$\qquad = 11.08 \text{ ft-kips}$

$I_e = I_s + I_c$

where

$$I_s = n A_{\text{eff}}(d - c)^2$$

$$c = \frac{a}{\beta}$$

$$A_{\text{eff}} = \frac{P_n + A_s f_y}{f_y}$$

$$I_c = \frac{bc^3}{3}$$

$$A_{\text{eff}} = \frac{24.3 \text{ kips} + (0.93 \text{ in.}^2)(40 \text{ ksi})}{40 \text{ ksi}}$$

$$= 1.54 \text{ in.}^2$$

$$I_s = \frac{(29{,}000 \text{ ksi})(1.54 \text{ in.}^2)(3.82 \text{ in.} - 1.32 \text{ in.}/0.85)^2}{1500 \text{ ksi}}$$

$$= 153.0 \text{ in.}^4$$

$$I_c = \frac{(36.7 \text{ in.})(1.32 \text{ in.}/0.85)^3}{3}$$

$$= 45.8 \text{ in.}^4$$

$$I_e = 153.0 + 45.8$$

$$= 198.8 \text{ in.}^4$$

Total applied moment:

$$M_u = \frac{q_u h^2}{8} + \frac{P_{u1} e}{2} + \left(\frac{P_{u1} + P_{u2}}{2}\right) \Delta$$

where

$$\Delta = \frac{5\phi M_n h^2}{48 E_m I_c}$$

$$= \frac{(5)(11.08 \text{ ft-kips})(17)^2(1728 \text{ in.}^3/1 \text{ ft}^3)}{(48)(1500)(198.8)}$$

$$= 1.93 \text{ in.}$$

$$M_u = \frac{(0.13)(17)^2}{8} + \frac{(16.5)(0.33)}{2} + \left(\frac{16.5 \text{ kips}}{2} + \frac{3.5 \text{ kips}}{2}\right)$$

$$\cdot (1.93) \left(\frac{1 \text{ ft}}{12 \text{ in.}}\right)$$

$$= 10.4 \text{ ft-kips} < \phi M_n = 11.1 \text{ ft-kips} \qquad \text{Therefore, O.K.}$$

$$\Delta\text{max} = \frac{M_u \Delta}{\phi M_n}$$

$$= \frac{(10.4)(1.93)}{11.1}$$

$$= 1.8 \text{ in.}$$

The maximum deflection is calculated at the factored moment value. The deflection at service loading could be calculated in a similar manner, but without the load factors.

Shear Wall Design

The concrete load applied to the walls is computed by determining the dead load of the roof, the weight of the wall being designed, and half of the weight of the walls perpendicular to the applied lateral load. Only one-half of the perpendicular wall weight is used because it is assumed that the weight of the lower half of the wall is carried directly to the foundation. This implies that only the weight of the upper half of the wall must be distributed by the diaphragm.

North and south walls:

$$W = (0.013 \text{ ksf})(40)(100) + (0.058 \text{ ksf}) \left(\frac{17}{2} + 3\right)(40)(2)$$

$$+ (0.058 \text{ ksf})(20)(100)$$

$$= 221.4 \text{ kips}$$

$$V = 0.186W$$

$$= 0.186(221.4)$$

$$= 41.2 \text{ kips}$$

$$V_u = 2.0V$$
$$= 2.0(41.2)$$
$$= 82.4 \text{ kips}$$

Shear strength (refer to Table 5.3 of Volume 1):

$$\frac{h}{d} = \frac{20}{100} = 0.2 < 0.25$$

$$V_m = \frac{V_u}{A} = \frac{82.4 \text{ kips}}{(62.4 \text{ in.}^2/\text{ft})(100 \text{ ft})}$$
$$= 0.013 \text{ ksi}$$

$$\phi V_m = \phi 4.0\sqrt{f'_m}$$
$$= (0.68)(4.0)\sqrt{1500 \text{ psi}}$$
$$= 0.105 \text{ ksi} > 0.013 \text{ ksi} \qquad \text{Therefore, O.K.}$$

West wall:

$$W = (0.013 \text{ ksf})(80)(50) + (0.058 \text{ ksf})\left(\frac{17}{2} + 3\right)(50)(2)$$
$$+ (0.058 \text{ ksf})(20)(80)$$
$$= 211.5 \text{ kips}$$

$$V = 0.186W$$
$$= 39.3 \text{ kips}$$

$$V_u = 2.0W$$
$$= 78.7 \text{ kips}$$

Shear strength:

$$\frac{h}{d} = \frac{20}{80} = 0.25$$

$$V_m = \frac{V_u}{A} = \frac{78.7 \text{ kips}}{(62.4 \text{ in.}^2/\text{ft})(80 \text{ ft})}$$
$$= 0.016 \text{ ksi}$$

$$\phi V_m = \phi(4.0)\sqrt{f'_m}$$
$$= 0.106 \text{ ksi} > 0.016 \text{ ksi} \qquad \text{Therefore, O.K.}$$

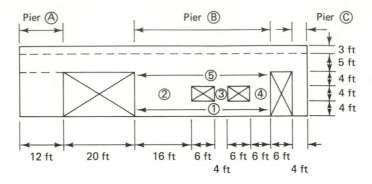

Figure 7.7 Model of East elevation for relative stiffness calculation

It is shown, for the walls with no openings (north, south, and west), that the masonry alone is sufficient to carry the applied load and that only the minimum wall steel is required. The plywood diaphragm is assumed to be incapable of distributing torsional shears. This implies flexible diaphragm behavior.

The distribution of the shear load to the elements forming the shear wall on the east side (Figure 7.7) is somewhat more involved than that used to find the design loads on the other shear walls. This occurs because the wall is composed of several elements of different widths, some with openings, and others without any openings. As discussed in Chapter 4, ignoring the openings would overestimate the capacity of the wall, while assuming that only the longer wall sections were available to resist the applied shear load would underestimate the capacity of the wall. The technique of distributing the shear loads to the various wall sections used in this case study is that presented in Section 4.6.

The contribution of each strip to the overall rigidity of the wall is computed. The rigidities of strips 2, 3, and 4 are added to give the rigidity at the window openings. The inverse of this rigidity is added to the inverse of the rigidities of strips 1 and 5 to give the relative displacement for pier *B*.

$$W = 211.5 \text{ kips} - (0.058 \text{ ksf})[(12)(20 + 6) + (4)(6)(2)]$$

$$= 190.6 \text{ kips}$$

$$V = 0.186W = 35.5 \text{ kips}$$

Strip 1 (fixed-fixed beam: $h/d = 4 \text{ ft}/38 \text{ ft} = 0.105$)

$$\Delta_{F-F} = \left(\frac{h}{d}\right)^3 + 3\left(\frac{h}{d}\right)$$

$$\Delta_1 = 1[(0.105)^3 + 3(0.105)]$$

$$= 0.317$$

$$R_1 = 3.155$$

Strip 2 (fixed-fixed beam: h/d = 4 ft/16 ft = 0.25)

$$\Delta_2 = 0.766$$

$$R_2 = 1.306$$

Strip 3 (fixed-fixed beam: h/d = 4 ft/4 ft = 1.0)

$$\Delta_3 = 4.0$$

$$R_3 = 0.25$$

Strip 4 (fixed-fixed beam: h/d = 4 ft/6 ft = 0.667)

$$\Delta_4 = 2.296$$

$$R_4 = 0.435$$

Strip 5 (fixed-fixed beam: h/d = 4 ft/38 ft = 0.237)

$$\Delta_5 = 0.317$$

$$R_5 = 3.155$$

Pier B

$$\Delta_B = \frac{1}{R_1} + \frac{1}{R_2 + R_3 + R_4} + \frac{1}{R_5}$$

$$= \frac{1}{3.155} + \frac{1}{1.306 + 0.25 + 0.435} + \frac{1}{3.155}$$

$$= 1.136$$

$$R_B = 0.880$$

Pier A (fixed-fixed beam: h/d = 12 ft/12 ft = 1.0)

$$\Delta_A = 4.0$$

$$R_A = 0.25$$

Pier C (fixed-fixed beam: h/d = 12 ft/4 ft = 3.0)

$$\Delta_C = 36.0$$

$$R_C = 0.028$$

It is seen that considering the opening in the wall represents a substantial reduction in the actual wall rigidity. Because the diaphragm is assumed to be flexible, the total amount of shear that must be carried by the east shear wall is equal to that carried by the west wall. The importance of this analysis is to determine how much shear is actually distributed to each element in the east wall. This will allow the designer to verify that the diaphragm to wall connection for each pier has sufficient capacity to carry the actual load, in addition to determining the design shear loads for each wall element.

Rigidity of wall with openings:

$$\Delta_T = 4 \left(\frac{5}{80}\right)^3 + 3 \left(\frac{5}{80}\right) + \frac{1}{0.880 + 0.25 + 0.028}$$

$$= 1.052$$

$$R_T = 0.950$$

Rigidity of wall without openings:

$$\Delta_{T1} = 4 \left(\frac{17}{80}\right)^3 + 3 \left(\frac{17}{80}\right)$$

$$= 0.676$$

$$R_{T1} = 1.48 > R_T = 0.950$$

The allocation of the shear load to the main piers (piers A, B, and C) is done first, after which the shears attributed to pier B are distributed to the component piers (piers 2, 3, and 4). This distribution is based on the relative rigidity of each wall. The wall elements with the greatest rigidity takes the largest shear load.

	Pier	R	$R/\Sigma R$	$V(R/\Sigma R)$
	A	0.250	0.216	7.7
	B	0.880	0.760	27.0
	C	0.028	0.024	0.9
$\Sigma R =$		1.158		35.6 kips
	2	1.306	0.656	17.7
	3	0.250	0.126	3.4
	4	0.435	0.218	5.9
$\Sigma R =$		1.991		27.0 kips

The factored shear load applied to each pier is compared to its shear capacity. It is shown that all elements have sufficient capacity to carry the shear load by relying only on the shear strength of the masonry. The flexural strength of the wall piers is also tabulated to determine the mode of failure of the piers.

Pier	$h/2d$[a]	Shear (in.^{2}) Area	$2.0V$	v_n[b] (ksi)	ϕV_n[c] (kips)	M_u[d] (ft-kips)
A	0.5	749	15.4	0.125	79.6	69.3
2	0.13	998	35.4	0.169	143.4	55.2
3	0.5	250	6.8	0.125	26.6	10.2
4	0.33	374	11.8	0.145	30.8	17.5
C	1.50	250	1.8	0.006	1.3	8.1

Note: $\phi V_n > V$ for all cases.
[a]$M/V_d = h/2d$.
[b]Interpolated values for v_n are from Table 5.3 of Volume 1.
[c]$\phi V_n = (0.85)(v_n)$(shear area).
[d]$M_u = 1.5(Vh/2)$.

The flexural capacity of each wall pier is compared to the cracking moment, to assure that sufficient jamb steel is provided to allow the wall sections to fail through yielding of the tension steel.

Pier A:

$$d = 144 - 4 = 140 \text{ in.} \qquad h = 12 \text{ ft}$$

$$M_{cr} = f_m S$$

$$= \frac{(0.1 \text{ ksi})(7.63 \text{ in.})(140 \text{ in.})^2(1 \text{ ft}/12 \text{ in.})}{6}$$

$$= 207.7 \text{ ft-kips}$$

Using Equation (5.11) try two No. 5 jamb steel bars:

$$\rho = \frac{A_s}{bd}$$

$$= \frac{(0.62 \text{ in.}^2)}{(7.63 \text{ in.})(140 \text{ in.})}$$

$$= 0.00058$$

$$\phi M_n = \phi \rho f_y bd^2 \left(1 - \frac{0.59 \rho f_y}{f'_m}\right)$$

$$= (0.8)(0.00058)(40 \text{ ksi})(7.63 \text{ in.})(140 \text{ in.})^2$$

$$\left[1 - \frac{(0.59)(0.00058)(40 \text{ ksi})}{1.5 \text{ ksi}}\right]\left(\frac{1 \text{ ft}}{12 \text{ in.}}\right)$$

$$= 229.4 \text{ ft-kips} > M_{cr} = 207.7 \text{ ft-kips} \qquad \text{Therefore, O.K.}$$

Moment at shear failure:

$$M_v = \frac{V_n h}{2}$$

$$= \frac{(79.6 \text{ kips})(12 \text{ ft})}{(2)(0.85)}$$

$$= 561.9 \text{ ft-kips} > \phi M_n = 229.4 \text{ ft-kips}$$

Therefore, flexure governs.
   Pier 2:

$$d = 192 - 4 = 188 \text{ in.} \qquad h = 4 \text{ ft}$$

$$M_{cr} = \frac{(0.1)(7.63)(188)^2}{(6)(12)}$$

$$= 374.5 \text{ ft-kips}$$

Use two No. 6 jamb steel bars:

$$\rho = \frac{0.88}{(7.63)(188)}$$

$$= 0.00061$$

$$\phi M_n = (0.8)(0.00061)(40)(7.63)(188)^2$$

$$\cdot \left[1 - \frac{(0.59)(0.00061)(40)}{1.5}\right]\left(\frac{1}{12}\right)$$

$$= 436.9 \text{ ft-kips} > M_{cr} = 374.5 \text{ ft-kips O.K.}$$

Moment at shear failure:

$$M_v = \frac{(143.4 \text{ kips})(4 \text{ ft})}{(0.85)(2)}$$

$$= 337.4 \text{ ft-kips} < \phi M_n = 436.9 \text{ ft-kips}$$

Therefore, shear governs.

Pier 3:

$$d = 48 - 4 = 44 \text{ in.} \quad h = 4 \text{ ft}$$

$$M_{cr} = \frac{(0.1)(7.63)(44)^2}{(6)(12)}$$

$$= 20.5 \text{ ft-kips}$$

Use one No. 5 jamb steel:

$$\rho = \frac{0.31}{(7.63)(44)}$$

$$= 0.00092$$

$$\phi M_n = (0.8)(0.00092)(40)(7.63)(44)^2$$

$$\cdot \left[1 - \frac{(0.59)(0.00092)(40)}{1.5} \right] \left(\frac{1}{12} \right)$$

$$= 35.8 \text{ ft-kips} > M_{cr} = 20.5 \text{ ft-kips O.K.}$$

Moment at shear failure:

$$M_v = \frac{(26.6)(4)}{(2)(0.85)}$$

$$= 62.6 \text{ ft-kips} > \phi M_n = 35.8 \text{ ft-kips}$$

Therefore, flexure governs.

Pier 4:

$$d = 72 - 4 = 68 \text{ in.} \quad h = 4 \text{ ft}$$

$$M_{cr} = \frac{(0.1)(7.63)(68)^2}{(6)(12)}$$

$$= 49.0 \text{ ft-kips}$$

Use one No. 5 jamb steel bar:

$$\rho = \frac{0.31}{(7.63)(68)}$$

$$= 0.0006$$

$$\phi M_n = (0.8)(0.0006)(40)(7.63)(68)^2$$

$$\cdot \left[1 - \frac{(0.59)(0.0006)(40)}{1.5} \right] \left(\frac{1 \text{ ft.}}{12 \text{ in.}} \right)$$

$$= 55.7 \text{ ft-kips} > M_{cr} = 49.0 \text{ ft-kips O.K.}$$

Moment at shear failure:

$$M_v = \frac{(30.8)(4)}{(2)(0.85)}$$

$$= 72.5 \text{ ft-kips} > \phi M_n = 55.7 \text{ ft-kips}$$

Therefore, flexure governs.

   Pier C:

$$d = 48 - 4 = 44 \text{ in.} \qquad h = 12 \text{ ft}$$

$$M_{cr} = 20.5 \text{ ft-kips}$$

Use one No. 5 jamb steel:

$$\phi M_n = 35.8 \text{ ft-kips} > M_{cr} = 20.5 \text{ ft-kips O.K.}$$

Moment at shear failure:

$$M_v = \frac{(1.3)(12)}{(2)(0.85)}$$

$$= 9.2 \text{ ft-kips} < \phi M_n = 35.8 \text{ ft-kips}$$

Therefore, shear governs.

The moment strength of each pier is compared to the applied moment caused by the ultimate shear loading. Assuming that sufficient shear capacity has been provided (i.e. $\phi V_n > V_u$), if the moment capacity of the wall pier is less than the moment caused by V_u flexure governs the failure of the wall. Piers A, 3, and 4 exhibit this kind of behavior. Piers 2 and C are governed by shear because the moment capacity of the piers is greater than the moment caused by V_u. If the small amount of flexural steel already in piers 2 and C were reduced to the point that flexure governed, the wall might fail at the cracking moment.

Chord Design

The applied moment on the diaphragm is calculated, assuming a uniformly applied load on a simply supported beam. The force in the chord is established by assuming that the internal moment arm is equal to the distance between the opposing walls, as presented in Section 4.5.

Try one No. 6 bar at north and south wall roof line.

$$W = \left[2\left(\frac{17 \text{ ft}}{2} + 3 \text{ ft}\right)(0.058 \text{ ksf}) + (80 \text{ ft})(0.013 \text{ ksf})\right](0.186)$$

$$= 0.442 \text{ klf}$$

$$M = \frac{wl^2}{8}$$

$$= \frac{(0.442)(100 \text{ ft})^2}{8}$$

$$= 552 \text{ ft-kips}$$

$$T_u = C_u = \frac{1.5M}{d}$$

$$= \frac{(1.5)(552 \text{ ft-kips})}{80 \text{ ft}}$$

$$= 10.4 \text{ kips}$$

$$A_s = \frac{T_u}{\phi f_y}$$

$$= \frac{10.4 \text{ kips}}{(0.8)(40 \text{ ksi})}$$

$$= 0.33 \text{ in.}^2 < 0.44 \text{ in.}^2$$

Thus the No. 6 bar at the north and south roof line is adequate for the tension load imposed on the chord by the diaphragm.

The compressive capacity of the chord must be evaluated as well. If it is assumed that only one horizontal course of masonry is grouted (i.e. the course with the chord steel), then the reduced nominal capacity, ϕP_n, can be determined using equation 5.19. In establishing the compressive capacity of the chord, it is probably reasonable to discount the contribution of the reinforcing steel $(A_s f_y)$ because it is not confined by lateral ties, but slenderness reductions need not be considered because of the bracing provided by the diaphragm. From equation 5.19,

$$\phi P_n = \phi \, 0.85 f_m' bd$$

$$= 0.65(0.85)(1.5)(7.63)(7.63)$$

$$= 48.3 \text{ kips} > 10.4 \text{ kips} \quad \text{O.K.}$$

A similar analysis is required for the chord in the east and west walls.

Try one No. 5 bar at east and west wall roof line.

$$W = 0.490 \text{ klf}$$

$$M = \frac{wl^2}{8}$$

$$= \frac{0.490(80)^2}{8}$$

$$= 392 \text{ ft-kips}$$

$$T_u = C_u = \frac{1.5M}{d}$$

$$= \frac{1.5\ (392)}{100}$$

$$= 5.88 \text{ kips}$$

$$A_s = \frac{5.88}{(0.8)(40)}$$

$$= 0.18 \text{ in.}^2 < 0.31 \text{ in.}^2$$

Use one No. 5 bar in east and west wall at diaphragm level.

It is clear from the analysis for the north-south chord that the single grouted course is adequate.

The connection between the diaphragm and the chord must be carefully detailed so that it can resist forces applied perpendicular and parallel to the connection. Many codes require that connections be designed to resist arbitrary minimum loads (e.g. 200 pounds per foot). These minimum forces must be compared to the applied design forces to determine which govern the design.

7.3 FOUR-STORY SHEAR WALL DESIGN CASE STUDY

This case study demonstrates the procedures involved in the analysis and design of a four-story shear wall. The details of the roof and floor framing system are not dealt with in this case study, nor are the effects of torsion-induced shear.

The shear wall is modeled as a ductile cantilever beam. The desired ductility is obtained by controlling the moment capacity of the shear wall. The designer attempts to provide shear capacity in the wall such that the lateral force equal to the shear strength is greater than the force corresponding to the moment capacity of the wall. In this way the wall fails in flexure rather than shear. This ideal behavior is not always possible because of the geometry of the low- to medium-rise shear walls.

The design procedure for this type of shear wall consists of first determining the base shear and lateral force distribution at each floor using a code-specified equivalent lateral load or an elastic response spectrum. The lateral load plus appropriate gravity loads are applied to the system using the loading combinations specified by the code to establish the design criteria for moment at the base of the wall. Flexural reinforcing steel is then selected to provide the required moment strength. Using the lateral loads developed above, the required shear strength of the wall is determined and vertical and horizontal shear reinforcing is provided. For the purposes of illustration, a load factor of 2.0 is used to establish the required shear strength. This load factor is higher than that used to figure the moment capacity so that the wall will fail in flexure. The last design step checks the axial forces on the boundary elements of the wall to establish the need for pilasters.

The design verification process then checks the capacity of the wall against the loads anticipated during a collapse-level earthquake by using an inelastic response spectrum analysis. The inelastic response spectrum, a modified form of the elastic response spectrum, accounts for the nonlinear behavior observed in structural systems loaded beyond their yield strength. The construction of an inelastic response spectrum is discussed both in Chapter 6 of Volume 1 and in Volume 3 of this series.

The use of an inelastic response spectrum in the design process requires some comment. Once the yield-level loads are established and the design completed, the performance of the system should be checked against a collapse-level earthquake (design verification). Although it is possible to design structural systems which would remain in a subyield state even during a so-called *maximum credible event*, it is usually not economical to design a building to do so. A collapse-level earthquake is defined to have a probability of being exceeded in 50 years of only 10 percent. Consequently, it is typically assumed that some yielding and inelastic response of the system occur when subjected to seismic loads such as these.

Loads generated by response spectrum analysis are, by definition, assumed to be ultimate loads. For this reason, *they are not factored*. If these loads were to be factored, it would imply uncertainty in addition to that already explicitly incorporated into the development of the response spectrum. Such layering of conservatism on top of conservatism is not justified when attempting to produce an economical structural design to perform in a safe manner during extreme ground motions.

In the case study at hand, a system ductility of 1.5 is assumed for an elastoplastic system with 5 percent damping. Higher values of ductility and damping are certainly justifiable. The spectral acceleration used is 0.63g, as compared to the effective acceleration generated from the UBC approach of 0.177g. If the UBC value is increased by a load factor of 1.5 [see Equation (5.3e)], the implied system ductility factor associated with the code-specified base shears may be determined [see page 100 of Volume 1].

$$\mu = \frac{0.63}{(1.5)(0.177)} = 2.37$$

Although more liberal than the value used in the case study, it appears to agree reasonably well with the chosen ductility factor of 1.5.

The loads used in the inelastic response spectrum analysis are used to simulate a system that is assumed to have yielded. This simulation takes the form of reducing the loads that would be expected in a collapse-level event to an equivalent load which may then be used to verify the fact that the system has sufficient capacity to withstand such an earthquake. The loads in an inelastic response spectrum are reduced from those produced by an elastic response spectrum with equivalent ground acceleration because inelastic deformation of the structure is permitted to occur. In the design example presented, the shear capacity of the wall is shown to be adequate to undergo a system ductility demand of at least 1.5 because the shear strength of the wall is greater than the base shear generated from the inelastic response spectrum analysis.

It may be noted that an elastic response spectrum analysis might have been used for the initial design of the wall in place of the UBC analysis. The earthquake ground motion in this case would have been based on a damage-level earthquake. The inelastic response spectrum analysis would have been performed in exactly the same manner, using the collapse level earthquake as the input ground motion.

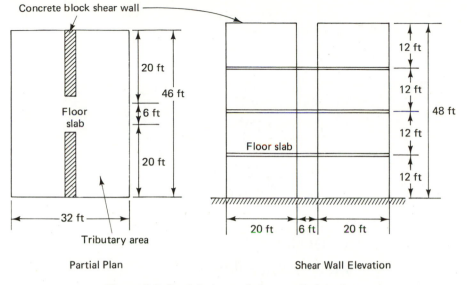

Figure 7.8 Partial plan and shear wall elevation

The shear wall in Figure 7.8 is to be constructed of 8-in. fully grouted concrete block. The ultimate strength of the block is $f'_m = 1.5$ ksi and Grade 40 steel ($f_y = 40$ ksi) will be used. Roof and floor dead loads are 75 psf and 100 psf, respectively. The roof live load is the same as that shown in Section 7.2. The floor live load is 50 psf and the wall dead load is 84 psf.

The lateral load is developed in the same manner as is used in Section 7.2. The value of *ZICKS* of 0.177 is derived.

Lateral load: from the 1982 *Uniform Building Code:*

$$V = ZICSKW$$

$$Z = 1.0 \text{ for zone 4}$$

$$I = 1.0 \text{ for commerical building}$$

$$K = 1.33 \text{ for box system}$$

$$T = \frac{0.05h}{\sqrt{D}}$$

$$= \frac{0.05(48 \text{ ft})}{\sqrt{46} \text{ ft}}$$

$$= 0.35 \text{ sec}$$

$$C = \frac{1}{15\sqrt{T}}$$

$$= \frac{1}{15\sqrt{0.35}}$$

$$= 0.11 < C_{max} = 0.12$$

$$S = 1 + \frac{T}{T_s} - 0.5 \left(\frac{T}{T_s}\right)^2 \qquad \text{assume that } T_s = 1.5 \text{ sec.}$$

$$= 1 + \frac{0.35}{1.5} - 0.5 \left(\frac{0.35}{1.5}\right)^2$$

$$= 1.2$$

$$CS = (0.11)(1.2)$$

$$= 0.133 < CS_{max} = 0.14$$

$$V = (1.0)(1.0)(0.133)(1.33)W$$

$$= 0.177W$$

$$W = [0.075 \text{ ksf} + (3)(0.100 \text{ ksf})](32 \text{ ft})(23 \text{ ft})$$

$$+ (0.084 \text{ ksf})(20 \text{ ft})(48 \text{ ft})$$

$$= 356.6 \text{ kips}$$

$$V = (0.177)(356.6 \text{ kips})$$

$$= 63.0 \text{ kips} \qquad \text{(for each wall)}$$

An initial check of the minimum reinforcing is performed to verify that this requirement is met by the design. It is shown, in the rest of the example, that this amount of steel is essentially sufficient to carry most all of the imposed loads.

Minimum reinforcing:

$$A_{s\,min} = (0.002)(12 \text{ in.})(8 \text{ in.})$$

$$= 0.19 \text{ in.}^2 \text{ per foot of wall}$$

Assume $\frac{2}{3}$ $A_{s\,min}$ as horizontal steel and $\frac{1}{3}$ $A_{s\,min}$ as vertical steel.

$$A_{sh} = \frac{2}{3}(0.19 \text{ in.}^2)$$

$$= 0.13 \text{ in.}^2/\text{ft}$$

$$A_{sv} = \tfrac{1}{3}(0.19 \text{ in.}^2)$$

$$= 0.06 \text{ in.}^2/\text{ft}$$

Therefore, use No. 5 bar at 32 in. on center for horizontal steel and No. 5 bar at 48 in. on center for vertical steel.

The stability of the shear wall system is checked by comparing the overturning moment produced by the applied lateral load with the dead-load resisting moment. The load distribution, shown in Figure 7.9, is used to calculate floor shears and overturning moments. (See Chapters 3 and 4 of Volume 1.)

Roof and floor loads:

$$P_r = (32 \text{ ft})(23 \text{ ft})(0.075 \text{ ksf}) + (20 \text{ ft})(6 \text{ ft})(0.084 \text{ ksf})$$

$$= 65.3 \text{ kips}$$

$$P_{fl} = (32 \text{ ft})(23 \text{ ft})(0.10 \text{ ksf}) + (20 \text{ ft})(12 \text{ ft})(0.084 \text{ ksf})$$

$$= 93.7 \text{ kips}$$

Dead-load resisting moment:

$$RM = [65.3 \text{ kips} + (3)(93.7 \text{ kips}) + (20 \text{ ft})(6 \text{ ft})(0.084 \text{ ksf})](10 \text{ ft})$$

$$= 3564.8 \text{ ft-kips}$$

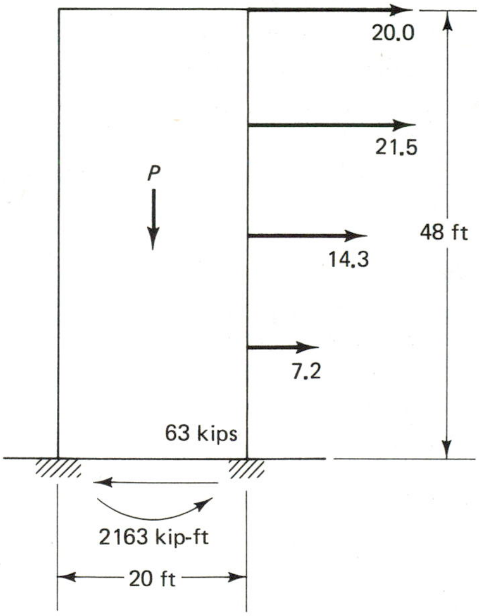

Figure 7.9 Model of shear wall loading

Overturning moment (OTM):

Floor	W_i (kips)	h_i (ft)	$W_i h_i$	F_i (kips)	OTM (ft-kips)
R	65.3	48	3134.4	20.0	–
4	93.7	36	3373.2	21.5	240.0
3	93.7	24	2248.8	14.3	738.0
2	93.7	12	1124.4	7.2	1407.6
1	–	–	–	–	2163.6
			$\Sigma W_i h_i$ = 9880.8	63.0	

Stability at ultimate moment:

$$M_u = 0.9(3564.8) - 1.5(2163.6)$$
$$= -37.1 \text{ ft-kips}$$

The negative overturning moment indicates that the foundation must provide restraint.

The cracking moment of the wall is established to determine the amount of jamb steel that must be provided to prevent a sudden tensile failure in the masonry as it reaches its tensile strength. A tensile strength of 0.10 ksi is assumed for the masonry. It is assumed, in this case, that there is no downward axial load.

The different loading combinations are also computed. These factored loads will be used to calculate the moment capacity of the wall.

Using Equation (3.14) to calculate the cracking moment, we obtain

$$M_{cr} = f_{mt}S$$

$$= \frac{(0.1 \text{ ksi})(7.63 \text{ in.})(240 \text{ in.})^2(1 \text{ ft}/12 \text{ in.})}{6}$$

$$= 610.4 \text{ ft-kips}$$

Try four No. 5 bars for jamb steel with d equal to 236 in. using Equation (5.11).

$$\rho = \frac{(0.31 \text{ in.}^2)(4)}{(7.63 \text{ in.})(236)}$$

$$= 0.00069$$

$$\phi M_n = (0.8)(0.00069)(40 \text{ ksi})(7.63)(236)^2$$

$$\cdot \left[1 - \frac{(0.59)(0.00069)(40 \text{ ksi})}{1.5 \text{ ksi}}\right]\left(\frac{1 \text{ ft.}}{12 \text{ in.}}\right)$$

$$= 771.9 \text{ ft-kips} > M_{cr} = 610.4 \text{ ft-kips} \quad \text{Therefore, OK.}$$

Use four No. 5 bars in first two cells at each end.

Using the amount of jamb steel determined from the analysis of the cracking moment, the flexural capacity of the wall is established for the two loading combinations. The wall is modeled as an eccentrically loaded column section as shown in Figure 7.10 and using the principles of force equilibrium and strain compatibility, the moment capacity of the wall is established. It is shown that the governing load combination is $0.9D - 1.5E$.

Load combinations:

A. $1.2D + 1.5E + 0.5L$

$$P_u = 1.2[65.3 \text{ kips} + (3)(93.7 \text{ kips})$$

$$+ (20 \text{ ft})(6 \text{ ft})(0.084 \text{ ksf})]$$

$$+ 1.5(0) + (0.5)[(8.8 \text{ kips}) + (3)(36.8 \text{ kips})]$$

$$= 487.4 \text{ kips}$$

$$M_u = 1.2(0) + 1.5(2163.6 \text{ ft-kips}) + 0.5(0)$$

$$= 3245.4 \text{ ft-kips}$$

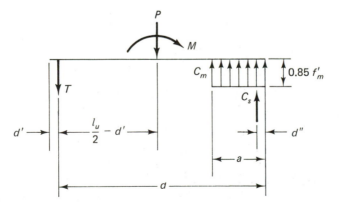

Figure 7.10 Force and moment equilibrium

B. $0.9D - 1.5E$

$$P_u = 0.9(346.4) - 1.5(0)$$

$$= 311.7 \text{ kips}$$

$$M_u = 0.9(0) - 1.5(2163.6)$$

$$= -3245.4 \text{ ft-kips}$$

Determine the moment capacity of the wall using load combination A

$$P = \frac{P_u}{\phi}$$

$$= \frac{487.4 \text{ kips}}{0.64}$$

$$= 761.5 \text{ kips}$$

$$P + T = C_m + C_s$$

$$P + f_y A_s = (0.85)(f'_m)(a)(b) + f'_s A'_s$$

$$\epsilon'_s = \frac{\epsilon_y (c - d'')}{d - c}$$

$$= \frac{\epsilon_y (a/0.85 - d'')}{d - a/0.85}$$

Rewriting the equilibrium equation to solve for a by using the quadratic formula gives us

$a^2 f'_m b - a/0.85(P + f_y A_s + f_y A'_s + 0.85^2 f'_m db) + d(P + f_y A_s)$
$+ f_y A'_s d'' = 0$

$a^2 (1.5 \text{ ksi})(7.63 \text{ in.}) - (a/0.85)[761.5 \text{ kips} + 40 \text{ ksi}(2)(1.24 \text{ in.}^2)$
$+ 0.85^2 (1.5 \text{ ksi})(236 \text{ in.})(7.63 \text{ in.})] + 236[761.5 \text{ kips}$
$+ (40 \text{ ksi})(1.24 \text{ in.}^2)] + (40 \text{ ksi})(1.24 \text{ in.}^2)(4 \text{ in.}) = 0$

$a^2 (11.44) - a(3308.4) + 191,618 = 0$

$$a = \frac{(3308.4) \pm \sqrt{(3308.4)^2 - 4(11.44)(191,618)}}{(2)(11.44)}$$

$$= 80.11 \text{ in.}$$

Find the strain in the compression steel:

$$\epsilon_s' = \frac{(40 \text{ ksi})(80.11 \text{ in.}/0.85 - 4 \text{ in.})}{(29,000 \text{ ksi})(236 \text{ in.} - 80.11 \text{ in.}/0.85)}$$

$$= 0.0009 < \epsilon_y = 0.00014 \qquad \text{Therefore, O.K.}$$

Check equilibrium:

$$P + f_y A_s - 0.85 f_m' ab - f_s' A_s' = 0$$

761.5 kips

+ (40 ksi)(1.24 in.²) - (0.85)(1.5 ksi)(80.11 in.)(7.63 in.)

- (0.0009)(29000 ksi)(1.24 in.) = 0

Find the moment capacity:

$$M_n = C_m \left(d - \frac{a}{2}\right) + C_s(d - d'') - P\left(\frac{l_u}{2} - d'\right)$$

$$= 779.3 \text{ kips} \left(236 \text{ in.} - \frac{80.11 \text{ in.}}{2}\right) + 31.6 \text{ kips } (236 \text{ in.} - 4 \text{ in.})$$

$$- 761.5 \text{ kips } (240 \text{ in.}/2 - 4 \text{ in.})$$

$$= 71,697 \text{ in.-kips} = 5975 \text{ ft-kips}$$

Find the shear load that causes the section to fail in flexure:

$$V = \frac{M_n \phi}{(2/3)h}$$

$$= \frac{(5974.8 \text{ ft-kips})(0.8)}{(2/3)(48 \text{ ft})}$$

$$= 149.4 \text{ kips}$$

Determine the moment capacity of the wall using load combination B.

$$P = \frac{311.7 \text{ kips}}{0.64}$$

$$= 487.1 \text{ kips}$$

$a^2(1.5 \text{ ksi})(7.63 \text{ in.}) - (a/0.85)[487.1 \text{ kips} + 40 \text{ ksi}(2)(1.24 \text{ in.}^2)$

$+ 0.85^2(1.5 \text{ ksi})(236 \text{ in.})(7.63 \text{ in.})] + 236 \text{ in.}[487.1 \text{ kips}$

$+ (40 \text{ ksi})(1.24 \text{ in.}^2)] + 40 \text{ ksi}(1.24 \text{ in.}^2)(4 \text{ in.}) = 0$

$$a^2(11.44) - a(2986) + 126{,}860 = 0$$

$$a = \frac{(2986) \pm \sqrt{(2986)^2 - 4(11.44)(126{,}860)}}{(2)(11.44)}$$

$$= 53.4 \text{ in.}$$

Find the strain in the compression steel:

$$\epsilon_s' = \frac{(40 \text{ ksi})(53.4 \text{ in.}/0.85 - 4 \text{ in.})}{(29{,}000 \text{ ksi})(236 \text{ in.} - 53.4 \text{ in.}/0.85)}$$

$$= 0.00047 < \epsilon_y = 0.0014$$

Check equilibrium:

$$P + f_y A_s - 0.85 f_m' ab - f_s' A_s' = 0$$

$$487.1 \text{ kips} + (40 \text{ ksi})(1.24 \text{ in.}^2)$$

$$- (0.85)(1.5 \text{ ksi})(53.4 \text{ in.})(7.63 \text{ in.})$$

$$- (0.00047)(29{,}000 \text{ ksi})(1.24 \text{ in.}^2) = 0$$

Find the moment capacity:

$$M_n = (519.5 \text{ kips})(236 \text{ in.} - 53.4 \text{ in.}/2) + 16.9 \text{ kips} (236 \text{ in.} - 4 \text{ in.})$$

$$- 487.1 \text{ kips} (240 \text{ in.}/2 - 4 \text{ in.})$$

$$= 56{,}149 \text{ kips in.-kips} = 4679 \text{ ft-kips}$$

Find the shear capacity that causes the section to fail in flexure:

$$V = \frac{(4679 \text{ ft-kips})(0.8)}{(2/3)(48 \text{ ft})}$$

$$= 117.0 \text{ kips}$$

Load combination B governs since it produces the smallest shear load, 117.0 kips, to cause the section to fail in flexure.

The capacity of the shear wall depends on the shear strength of the masonry and the steel. It is shown in the example that the masonry, acting alone, does not have sufficient capacity to carry the total shear load.

The effective shear area of the wall must be computed to establish the ultimate shear stress. The wall is assumed to be fully grouted, so the effective shear area is taken as 91.5 in.^{2}/ft. The effective length of the wall is assumed to be 80 percent of the total length.

$$V_u = 2.0V$$

$$= 2.0(63 \text{ kips})$$

$$= 126 \text{ kips}$$

Shear wall area:

$$A = (0.80)(20 \text{ ft})(91.5 \text{ in.}^2/\text{ft})$$

$$= 1464 \text{ in.}^2$$

$$v_u = \frac{V_u}{A}$$

$$= \frac{126 \text{ kips}}{1464 \text{ in.}^2}$$

$$= 0.086 \text{ ksi}$$

Using Table 5.3 from Volume 1, the ultimate masonry shear strength is calculated.

$$\frac{M_u}{V_u d} = \frac{(2.0)(2163.6 \text{ kip-ft})}{(126 \text{ kips})(0.8)(20 \text{ ft})}$$

$$= 2.15 > 1.0$$

Therefore,

$$v_m = 1.7 \sqrt{f'_m}$$

$$= 1.7\sqrt{1500} \text{ psi}$$

$$= 0.066 \text{ ksi} < v_u = 0.086 \text{ ksi} \qquad \text{Therefore, N.G.}$$

Using Table 5.3 from Volume 1, the ultimate allowable shear stress of masonry and reinforcement is calculated.

$$\frac{M_u}{V_u d} = 2.15 > 1.0$$

Therefore,

$$v_m = 2.5 \sqrt{f'_m}$$

$$= 2.5 \sqrt{1500} \text{ psi}$$

$$= 0.097 \text{ ksi} > v_u = 0.086 \qquad \text{Therefore, O.K.}$$

The shear capacity of the wall with No. 5 bars at 32 in. on center is

$$\phi V_n = \phi V_m + \phi V_s$$

$$= \phi \left[v_m t d + \frac{A_v f_y d}{s} \right]$$

$$= 0.68 \left[(0.066 \text{ ksi})(0.8)(91.5 \text{ in.}^2/\text{ft})(20 \text{ ft}) \right.$$

$$\left. + \frac{(0.31 \text{ in.}^2)(40 \text{ ksi})(0.8)(20 \text{ ft})(12 \text{ in.}/\text{ft})}{32 \text{ in.}} \right]$$

$$= 116.3 \text{ kips} < V_u = 126 \text{ kips}$$

Therefore, an increase in horizontal wall steel is required. Try No. 5 bars at 24 in. on center.

$$\phi V_n = 0.68 \left[(0.0635 \text{ ksi})(0.8)(91.5 \text{ in.}^2/\text{ft})(20 \text{ ft}) \right.$$

$$\left. + \frac{(0.31 \text{ in.}^2)(40 \text{ ksi})(0.8)(20 \text{ ft})(12 \text{ in.}/\text{ft})}{24 \text{ in.}} \right]$$

$$= 130.7 \text{ kips} > V_u = 126 \text{ kips} \qquad \text{Therefore, O.K.}$$

Use No. 5 horizontal rebar at 24 in. on center.

A comparison is made to determine the failure mode of the wall. The shear load to fail the wall in flexure is 117.0 kips. The shear capacity of the wall is 130.7 kips. Since different loading factors are used to determine these shear values, an adjustment is required to make a comparison.

$$V_f = 117.0 \text{ kips}$$

$$V_n = 130.7 \text{ kips}$$

$$V_f = 117.0 \text{ kips} \left(\frac{2.0}{1.5} \right) = 156.0 \text{ kips} > V_n = 130.7 \text{ kips}$$

Therefore, the wall will fail in shear before failing in flexure and to achieve the more desirable failure mode (i.e. flexure), the shear capacity of the wall should be increased.

The requirements contained in the ACI document 318-77, Sections A8.4 and A8.5, are intended to ensure ductility in the wall section. The condition of flexure with light axial load is considered

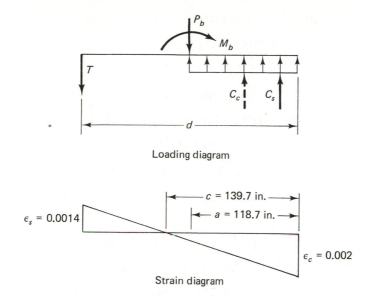

Loading diagram

Strain diagram

Figure 7.11 Loading and strain diagrams

in Section A8.5. In a parallel fashion we may determine if pilasters are required at the ends of the masonry shear wall. The balanced axial load is determined using Figure 7.11.

$$c = \frac{\epsilon_m d}{\epsilon_s + \epsilon_m}$$

$$= \frac{0.002(236 \text{ in.})}{0.0014 + 0.002}$$

$$= 139.7 \text{ in.}$$

$$a = \beta c$$

$$= 0.85(139.7 \text{ in.})$$

$$= 118.7 \text{ in.}$$

$$P_b = C_c + C_s - T$$

$$= C_c \quad \text{since } T = C_s$$

$$= 0.85 f'_m A_m$$

$$= 0.85(1.5 \text{ ksi})(118.7 \text{ in.} \times 7.63 \text{ in.})$$

$$= 1154.7 \text{ kips}$$

$$0.40\phi P_b = 0.4(0.64)(1154.7 \text{ kips})$$

$$= 295.6 \text{ kips}$$

$$P_D = 65.3 \text{ kips} + 3(93.7 \text{ kips})$$

$$= 346.4 \text{ kips} \quad (\text{dead load})$$

$$P_L = 3(0.05 \text{ kip/ft}^2)(32 \text{ ft})(23 \text{ ft})$$

$$+ (0.012 \text{ kip/ft}^2)(32 \text{ ft})(23 \text{ ft})$$

$$= 119.2 \text{ kips} \quad (\text{live load})$$

$$P_E = \frac{1.5(2163.6 \text{ ft-kips})}{(0.8)(20 \text{ ft})}$$

$$= 202.8 \text{ kips} \quad (\text{lateral load})$$

$$P_e = 1.2P_D + 1.5P_E + 0.5P_L$$

$$= 1.2(346.4) + 1.5(202.8) + 0.5(119.2)$$

$$= 779.5 \text{ kips} > 0.4\phi P_b = 295.6 \text{ kips}$$

Therefore, pilasters are required but their design is not shown as part of this case study.

The peak ground acceleration and peak ground velocity for a collapse-level earthquake are obtained from ATC-3 for Zone 7, which are $0.4g$ and 0.12 in./sec, respectively.

The inelastic response spectrum is established, (Figure 7.12), using the procedure outlined in Reference 6.1 of Volume 1. A coefficient of damping of 5 percent is used for an elastoplastic system. A ductility of 1.5 is assumed.

From Reference 6.1 of Volume 1, using $\beta = 5\%$ and $\mu = 1.5$,

$$S_a = \psi_u Y_a$$

$$\psi_u = 1.57 \quad \text{for the acceleration region}$$

Therefore,

$$S_a = (1.57)(0.4g)$$

$$= 0.63g$$

From Reference 6.1 of Volume 1, using $\beta = 5$ percent and $\mu = 1.5$,

$$S_v = \psi_u Y_v$$

$$\psi_u = 0.966 \quad \text{for the velocity region}$$

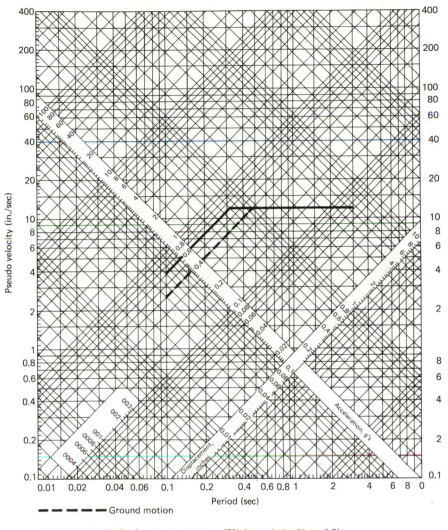

- - - - - Ground motion

Inelastic response spectrum (5% damped; ductility = 1.5)

Figure 7.12 Inelastic response spectrum

Therefore,

$$S_v = (0.966)(12 \text{ in./sec})$$

$$= 11.59 \text{ in./sec}$$

The distribution of lateral forces in a multistory building using a response spectrum analysis is discussed in Chapter 5 of Volume 1.

In the fundamental mode the vibration of a four-story shear wall

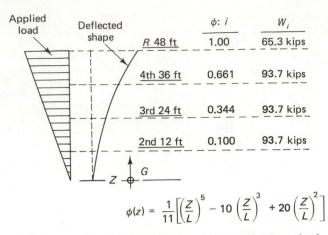

$$\phi(z) = \frac{1}{11}\left[\left(\frac{z}{L}\right)^5 - 10\left(\frac{z}{L}\right)^3 + 20\left(\frac{z}{L}\right)^2\right]$$

Figure 7.13 Shape function of cantilever beam with triangular loading

is assumed to behave similarly to a cantilever beam with a triangular load. This shape function is used to compute the values of L and M^* (see Figure 7.13).

$$f_i = \frac{LS_a m_i \phi_i}{M^*}$$

where $L = \Sigma\, m_i \phi_i$

$\quad\ M^* = \Sigma\, m_i \phi_i^2$

The base shear is expressed as

$$V = \Sigma\, f_i = \frac{LS_a\, \Sigma\, m_i \phi_i}{M^*}$$

$$= \frac{L^2 S_a}{M^*}$$

$$L = \frac{1}{g}\, \Sigma\, W_i \phi_i$$

$$= \frac{1}{g}\, [(65.3)(1.00) + 93.7(0.661 + 0.344 + 0.100)]$$

$$= \frac{1}{g}\, [168.8\ \text{kips}]$$

$$M^* = \frac{1}{g}\, \Sigma\, W_i \phi_i^2$$

$$= \frac{1}{g} \left[65.3(1.00)^2 + 93.7(0.661^2 + 0.344^2 + 0.100^2) \right]$$

$$= \frac{1}{g} \left[118.3 \text{ kips} \right]$$

The fundamental elastic period falls in the velocity controlled region of the response spectrum. The spectral velocity must then be transformed to a psuedo-acceleration, for use in the response spectrum analysis.

$$T_n = 0.35 \text{ sec}$$

$$S_v = 11.59 \text{ in./sec}$$

Therefore,

$$S_a = \left(\frac{2\pi}{T_n} \right) S_v$$

$$= \left(\frac{2\pi}{0.35 \text{ sec}} \right) (11.59 \text{ in./sec})$$

$$= 208 \text{ in./sec}^2$$

$$= 0.538g$$

The wall base shear:

$$V = \frac{L^2 S_a}{M^*}$$

$$= \frac{(1/g)(168.8 \text{ kips})^2 (0.538g)}{118.3 \text{ kips}}$$

$$= 129.6 \text{ kips } (=0.37W)$$

The inelastic building period is used for displacement calculations.

$$T_n' = \sqrt{\mu} \, T_n$$

$$= \sqrt{1.5} \, (0.35 \text{ sec})$$

$$= 0.43 \text{ sec}$$

Level	F_i (kips)	q (in.)	OTM (ft-kips)
R	50.2	1.70	—
4	47.5	1.12	602.4
3	24.7	0.58	1774.8
2	7.2	0.17	3243.6
6	—	—	4798.8
	129.6 kips		

$$V = 129.6 \text{ kips} < \phi V_n = 130.7 \text{ kips}$$

The base shear, from the inelastic response spectrum, is less than the shear capacity of the wall. Thus, a collapse-level earthquake, requiring a ductility demand of 1.5, seems to be within the capacity of the wall.

The maximum interstory drift is shown to be smaller than the allowable drift.

$$\text{Maximum } \Delta = 1.70 \text{ in.} - 1.12 \text{ in.}$$

$$= 0.58 \text{ in.}$$

$$\text{Allowable } \Delta = 0.005 h_x$$

$$= (0.005)(144 \text{ in.})$$

$$= 0.72 \text{ in.}$$

A stability check is made and shows the overturning moment is larger than the resisting moment. Therefore, the foundation must provide restraint.

$$\text{RM} - \text{OTM} = 0.9(3564.8 \text{ ft-kips}) - 4798.8 \text{ ft}$$

$$= -1590.5 \text{ ft-kips}$$

The moment strength of the wall with four No. 5 bars in the first two cells at each end of the wall is compared to the required moment (M_u = OTM).

$$\phi M_n = 0.8(4679 \text{ ft-kips})$$

$$= 3743 \text{ ft-kips} < 4.798.8 \text{ ft-kips}$$

Additional flexural steel should be added to provide sufficient flexural capacity. This requires an additional iteration through the design process and the steps are identical to those shown in this case study, but the repeated calculations will not be shown. The addition of the boundary elements, shown to be required earlier in the case study, may provide sufficient flexural capacity without significantly increasing the amount of reinforcing steel required.

APPENDIX A

CALIBRATION OF STRENGTH REDUCTION FACTORS FOR CONCRETE MASONRY

A.1 GENERAL

Recent work has applied the concepts of uncertainty analysis in order to establish a set of values for load factors and loading combinations which are to be used in the design of all engineering materials [A.1]. The load factor values adopted by the American National Standards Institute [A.2] are intended to be used as part of a probability-based limit state design (PBLSD) methodology. The load factors and loading combinations were selected in a manner which achieves consistent levels of structural reliability under various loading conditions involving dead, live, wind, snow, and earthquake loads when used with compatible strength reduction factors. Values for strength reduction factors for concrete masonry using a PBLSD methodology must now be developed and these new strength reduction factors must be developed so that the final design results in structural members which have consistent levels of reliability [A.3].

The values of the strength reduction factors published in the American Concrete Institute Code (ACI 318-77) [A.4] are the result of subjective evaluations regarding the probability of structural failure in concrete members. Research has shown that these strength reduction factors result in varying degrees of structural reliability under different loading combinations because they do not explicitly consider the uncertainty associated with both the load and resistance sides of the design equation [A.5]. In addition to these empirically derived strength reduction factors, the load factors used in conjunction with ACI 318-77 are also not based on an explicit examination of the uncertainty associated with the load effects.

Despite the variation in structural reliability which results from a combination of the current nonprobabilistically derived load and strength reduction factors, the number of structural failures involving concrete members is quite small. This suggests that it would be reasonable to identify the mean values of reliability for the different limit states according to present code formulations and to calibrate the new strength reduction factors to match these mean levels of reliability. Such work has been done in concrete [A.5] and a set of revised strength reduction factors has been developed.

In this Appendix we develop a set of strength reduction factors which are calibrated to the strength reduction factors in reference [A.5]. This set of strength reduction factors is developed by considering the relative uncertainty in limit states and loading states between the two materials, concrete and concrete masonry, which are involved in the calibration [A.6].

A.2 RELIABILITY EQUIVALENCE FORMULATION

A measure of structural reliability is the reliability index, β. For a given limit state, we will develop a value of the strength reduction factor for concrete masonry (material A) using the existing strength reduction factors for concrete (material B). We will develop this value in such a way that the value of the reliability indices for the two materials are equal:

$$\beta_A = \beta_B \qquad (A.1)$$

The first step in this development is the derivation of an expression that relates the strength reduction factor for material A, ϕ_A, to the strength reduction factor for material B, ϕ_B.

The reliability index is a function of the statistics of the safety margin, F:

$$\beta = \frac{\overline{F}}{\sigma_F} \qquad (A.2)$$

where $\overline{F}$ = mean value of the safety margin

σ_F = standard deviation of the safety margin

The safety margin relates the effects of load, U, and structural resistance, R.   One definition of the safety margin is $F = R - U$. However, in the context of the present development, a formulation of F which is more convenient is found on page 1336 of reference [A.7].

$$F = \ln \left(\frac{R}{U} \right) \tag{A.3}$$

In this case, failure would be defined when $(R/U) \leqslant 1$ or $\ln (R/U) \leqslant 0$.

Assuming a linear statistical model [A.8], the mean, $\overline{F}$, and standard deviation, σ_F, of the safety margin are

$$\overline{F} = \ln \left(\frac{\overline{R}}{\overline{U}} \right) \tag{A.4a}$$

and

$$\sigma_F = (V_R^2 + V_U^2)^{0.5} \tag{A.4b}$$

where V_R and V_U = coefficient of variation of R and U, respectively = $\sigma_R/\overline{R}$, $\sigma_u/\overline{U}$. It then follows that the reliability index may be redefined in terms of Equations (A.4a) and (A.4b).

$$\beta = \frac{\ln (\overline{R}/\overline{U})}{(V_R^2 + V_U^2)^{0.5}}$$

From Equations (A.1) and (A.5) the reliability indices for the two materials, A and B, may be equated as shown below if it is assured that the load effects are equal.   This is assurred if the same load factors are used as proposed by ANSI. Thus,

$$\beta_A = \beta_B$$

and

$$\ln \left(\frac{\overline{R_A}}{\overline{U}} \right) \Big/ (V_{R_A}^2 + V_U^2)^{0.5} = \ln \left(\frac{\overline{R_A}}{\overline{U}} \right) \Big/ (V_{RB}^2 + V_U^2)^{0.5} \tag{A.6}$$

Rearranging Equation (A.6) gives us

$$\ln \left(\frac{\overline{R_A}}{\overline{U}} \right) (V_{RB}^2 + V_U^2)^{0.5} = \ln \left(\frac{\overline{R_B}}{\overline{U}} \right) (V_{R_A}^2 + V_U^2)^{0.5} \tag{A.7}$$

and taking the inverse of the natural log of both sides yields

$$\frac{\overline{R_A}}{\overline{U}} \exp (V_{RB}^2 + V_U^2)^{0.5} = \left(\frac{\overline{R_B}}{\overline{U}}\right) \exp (V_{RA}^2 + V_U^2)^{0.5} \qquad (A.8)$$

Multiplying both sides of Equation (A.8) by $\overline{U}$ and rearranging terms, Equation (A.9) is obtained:

$$\frac{\overline{R_A}}{\overline{R_B}} = \exp [(V_{RA}^2 + V_U^2)^{0.5} - (V_{RB}^2 + V_U^2)^{0.5}] \qquad (A.9)$$

A.3 STRENGTH REDUCTION FACTOR DEVELOPMENT

Structural designers rely on design equations to calculate the nominal strength of a member, R_n. If a series of representative structural sections is considered, it is possible to determine the statistical properties associated with the ratio of the mean structural resistance compared to the nominal structural resistance (i.e., $\overline{R}/R_n$).

The mean resistance, $\overline{R}$, may be thought of as being proportional to the nominal resistance, and may be expressed as

$$\overline{R} = CR_n \qquad (A.10)$$

where C is the constant of proportionality.

Equation (A.9) may be rewritten

$$\frac{\overline{R_A}}{\overline{R_B}} = \frac{C_A R_{nA}}{C_B R_{nB}}$$

$$= \exp [(V_{RA}^2 + V_U^2)^{0.5} - (V_{RB}^2 + V_U^2)^{0.5}] \qquad (A.11)$$

or

$$\frac{R_{nA}}{R_{nB}} = \frac{C_B}{C_A} \exp [(V_{RA}^2 + V_u^2)^{0.5} - (V_{RB}^2 + V_U^2)^{0.5}] \qquad (A.12)$$

The design equation in a limit state design methodology attempts to ensure that the reduced nominal resistance, ϕR_n, is greater than the factored demand, U. Thus, the minimal design condition occurs when the two terms are equal:

$$\phi R_n = U \qquad (A.13)$$

Replacing the left-hand side of Equation (A.13) with equivalent expressions for materials A and B, we obtain

$$\phi_A R_{nA} = \phi_B R_{nB}$$

or, alternatively,

$$\frac{R_{nA}}{R_{nB}} = \frac{\phi_B}{\phi_A} \qquad (A.14)$$

Combining Equations (A.12) and (A.14), it follows that

$$\frac{R_{nA}}{R_{nB}} = \frac{\phi_B}{\phi_A}$$

$$= \frac{C_B}{C_A} \exp\left[(V_{RA}^2 + V_U^2)^{0.5} - (V_{RB}^2 + V_U^2)^{0.5}\right] \qquad (A.15)$$

and solving for ϕ_A, we obtain

$$\phi_A = \phi_B \left(\frac{C_A}{C_B}\right) \exp\left\{-\left[(V_{RA}^2 + V_U^2)^{0.5} - (V_{RB}^2 + V_U^2)^{0.5}\right]\right\} \qquad (A.16)$$

Equation (A.16) makes it possible to calibrate a new set of strength reduction factors, ϕ_A, with an existing set, ϕ_B, provided that the other terms on the right-hand side are known.

An examination of the terms in Equation (A.16) will shed some light on the relative contributions of the variability in the material resistances and load effects upon changes in the magnitude of ϕ_A. It is clear that any changes resulting in ϕ_A from a change in ϕ_B or C_A/C_B are of a linear nature. The changes in ϕ_A caused by a change in the exponential term are of more interest.

As the coefficient of variation of the load effect, V_u, increases in magnitude, the value of ϕ_A increases for constant values of V_A and V_B. This implies that the relative variation in material strengths becomes less important as the variability of the load increases. In fact, at the limit as V_U, approaches infinity, the exponential term converges to unity, and assuming that $C_A/C_B = 1.0$, ϕ_A equals ϕ_B.

If the coefficient of variation of material A is less than that of material B, it means that there is less uncertainty associated with the strength of material A. This implies that the value of ϕ_A should be greater than ϕ_B. An examination of Equation (A.16) shows this to be true if it is assumed that $C_A/C_B = 1.0$. Conversely, if $V_A > V_B$, then $\phi_A < \phi_B$. The trend is clear: the material with less uncertainty is associated with the larger strength reduction factor.

TABLE A.1 Strength Reduction Factors for Reinforced
Concrete [A.5]

Limit State	ϕ_c
Flexure	
Low steel content[a]	0.85
High steel content[b]	0.75
Axial tension and tension plus flexure	0.85
Axial compression and compression plus flexure	0.65
Shear	0.70
Bearing on concrete	0.60

[a]Steel ratio, ρ, not greater than 60 percent of steel ratio at balanced condition, ρ_b.

[b]Steel ratio, ρ, greater than 60 percent of steel ratio at balanced condition, ρ_b.

A.4 CONCRETE MASONRY STRENGTH REDUCTION FACTOR CALIBRATION

Strength reduction factors for reinforced concrete, ϕ_c, are shown in Table A.1 [A.5].

A study of the information available for the variation in loading effects shows that in Table A.2 there is a wide range of values for V_U for dead, live, wind, snow, and seismic effects. It seems reasonable to evaluate the value of the strength reduction factor for concrete masonry, ϕ_m, using the different values of V_U and then examine the range of the results. It will be shown that there is only a small variation in the value of ϕ_m in spite of the wide variability of V_U as contained in Table A.2. The mean value of ϕ_m considering the different load types will be used to represent the strength reduction factor for concrete masonry for the failure mode under consideration.

The coefficients of variation for the resistance ratios, V_c and V_m,

TABLE A.2 Coefficient of Variation for Load
Effects, V_U [A.1]

Load type	V_U
Dead	0.10
Live	0.25
Wind	0.37
Snow	0.26
Earthquake	1.38

TABLE A.3 Resistance Ratio Statistics $\bar{R}/R_n$ [A.1]

Failure mode	Concrete		Concrete Masonry	
	$\dfrac{\bar{R}_c}{R_{nc}}$	V_c	$\dfrac{\bar{R}_m}{R_{nm}}$	V_m
Flexure	1.07	0.125	—	0.200
Tension	1.00	0.115	—	0.240
Compression	1.03	0.156	—	0.240
Shear	1.01	0.190	—	0.240
Bearing	—	0.115	—	0.240

are shown in Table A.3. These values were obtained from the appendices of reference [A.1] by averaging the different resistance categories such as flexure with Grade 60 steel or flexure with Grade 40 steel, and so on. There were no values of V_m for shear, tension, or bearing presented in reference [A.1]. Consequently a conservative value was assumed equal to that of masonry in compression. A con-

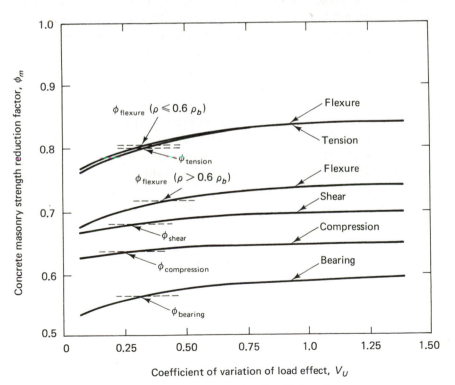

Figure A.1 Strength reduction factor, ϕ_m

TABLE A.4   Strength Reduction Factors for Concrete Masonry

Failure Mode	$\overline{\phi}_m$ (calculated)	ϕ_m (recommended design value) [A.6]
Flexure		
Low steel content[a]	0.801	0.80
High steel content[b]	0.717	0.70
Tension or tension plus flexure	0.798	0.80
Compression or compression plus flexure	0.645	0.65
Shear	0.680	0.65
Bearing on concrete masonry	0.564	0.55

[a]Steel ratio, ρ, not greater than 60 percent of steel ratio at balanced condition, ρ_b.

[b]Steel ratio, ρ, greater than 60 percent of steel ratio at balanced condition, ρ_b.

servative assumption was made in the case of bearing (high V_m and low V_c) because published statistics for both materials seem lacking in current literature.

There is presently no information regarding the value of C_m from Equation (A.9) because there is really no widespread use of strength design for concrete masonry members. This makes it difficult to evaluate the ratio of C_m/C_c. The ideal case occurs when $C_m/C_c = 1.0$. However, the results in this calibration may be easily modified once the actual values of C_m are obtained.

Figure A.1 plots the variation in ϕ_m against the coefficient of variation in the load effect V_U. It can be seen that there is only a small change in the strength reduction factor over the entire range of V_U. The mean value, $\overline{\phi}_m$, and the recommended design value of ϕ_m for each of the failure modes are shown in Table A.4.

APPENDIX B

REFERENCES

1.1 Uniform Building Code, 1982 Edition. Whittier, CA: International Conference of Building Officials, 1982.

1.2 ACI Committee 531, *Building Code Requirements for Concrete Masonry Structures*, ACI 531-81. Detroit, MI: American Concrete Institute, 1981.

1.3 "Building Code Requirements for Minimum Design Loads in Buildings and Other Structures," ANSI A58.1-1982. New York: American National Standards Institute, 1982.

1.4 "Strength Design of Concrete Masonry Buildings," Los Angeles: Englekirk and Hart Consulting Engineers, Inc., February 1983.

2.1 M. Priestly, "Ductility of Unconfined and Confined Concrete Masonry Shear Walls," *TMS Journal*, The Masonry Society, July–December 1981, T28–T39.

2.2 D. C. Kent and R. Park, "Flexural Members with Confined Concrete." *Journal of the Structural Division*, ASCE, 97:ST7, 1969-1990, July 1971.

2.3 *Uniform Building Code*, 1982 Edition. Whittier, CA: International Conference of Building Officials, 1982.

2.4 R. G. Drysdale and A. A. Hamid, "Behavior of Concrete Block Masonry under Axial Compression." *ACI Journal*, June 1979, 707–721.

2.5 D. E. Allen, "Statistical Study of the Mechanical Properties of Reinforcing Bars," Building Research Note 85. Ottawa, Ontario: National Research Council of Canada, April 1972.

2.6   ACI Committee 531, *Building Code Requirements for Concrete Masonry Structures*, ACI 531-81. Detroit, MI: American Concrete Institute, 1981.

2.7   G. A. Hegemier, G. Krishnamoorthy, R. O. Nunn, and T. V. Moarthy, "Prism Tests for the Compressive Strength of Concrete Masonry." San Diego, CA: University of California, 1977.

2.8   *Passive Solar Masonry Handbook.* Sacramento, CA: Concrete Masonry Association of California and Nevada, 1981.

2.9   *ASHRAE Handbook of Fundamentals.* New York: American Society of Heating, Refrigeration, and Airconditioning Engineers, 1967.

2.10   *Concrete Masonry Design Manual.* Sacramento, CA: Concrete Masonry Association of California and Nevada, 1981, p. 113.

2.11   "Energy Conservation Standards for New Residential Buildings." Sacramento, CA: California Energy Commission, 1982.

2.12   "Grouting for Concrete Masonry Walls." Herndon, VA: National Concrete Masonry Association, 1979.

2.13   "Manual of Standard Practice," Chicago, IL: Concrete Reinforcing Steel Institute, 23rd Edition, 1981.

2.14   B. Wilcox, A. Gummerlock, and R. Mitchell, "California Energy Code: Credit for Massive Exterior Walls." *Progress in Passive Solar Energy Systems*, Vol. 7, New York: American Solar Energy Society, 1982.

3.1   M. Priestly, "Ductility of Unconfined and Confined Concrete Masonry Shear Walls." *TMS Journal*, The Masonry Society, July–December 1981, T28–T39.

3.2   D. Ramaley and D. McHenry, "Stress–Strain Curves for Concrete Strained beyond Ultimate Load," Lab Report No. SP-12. Denver, CO: U.S. Bureau of Reclamation, 1947.

3.3   "The Design of Reinforced Masonry," DZ4210, Part B. Wellington, N.Z.: SANZ, 1980.

3.4   G. Winter and A. H. Nilson, *Design of Concrete Structures.* New York: McGraw-Hill, Ninth Edition, 1979.

3.5   R. Park and T. Paulay, *Reinforced Concrete Structures.* New York: Wiley, 1975.

3.6   C. S. Whitney, "Design of Reinforced Concrete Members under Flexure or Combined Flexure and Direct Compression." *ACI Journal*, March–April 1937, 33: 483–498.

3.7   H. P. J. Taylor, "Further Tests to Determine Shear Stresses in Reinforced Concrete Beams," TRA 438. London: Cement and Concrete Association, February 1970, p. 27.

3.8   S. Chen, P. A. Hidalgo, R. L. Mayes, R. W. Clough, and H. D. McNiven, "Cyclic Loading Tests of Masonry Shear Piers," Volume II, UBC/EERC-78/28. Berkeley: University of California, 1978.

3.9   B. Bresler and J. G. MacGregor, "Review of Concrete Beams Failing in Shear." *Journal of the Structural Division*, ASCE, 93: ST2, 343–372, February 1967.

3.10 C. K. Wang and C. G. Salmon, *Reinforced Concrete Design*, 3rd ed. New York: Harper & Row, 1979.

3.11 ACI-ASCE Committee 326, "Shear and Diagonal Tension," *ACI Journal*, January, February, and March 1962, 59:1–30, 277–334, 352–396.

3.12 M. J. Haddadin, S. Hong, and A. H. Mattock, "Stirrup Effectiveness in Reinforced Concrete Beams with Axial Force." *Journal of the Structural Division*, ASCE, September 1971, 97:2277–2297.

3.13 A. H. Mattock and N. M. Hawkins, "Research on Shear Transfer in Reinforced Concrete." *PCI Journal*, March–April 1972, 17:55–75.

3.14 European Committee on Reinforced Concrete (CEB), *International Recommendations for the Design and Construction of Concrete Structures*, Appendix 3, first ed. London: Cement and Concrete Association, 1970.

3.15 ACI Committee 318, *Building Code Requirements for Reinforced Concrete*, ACI 318-83. Detroit, MI: American Concrete Institute, 1983.

3.16 R. E. Englekirk and G. C. Hart, "Concrete Block Shear Walls: Strength Design for Flexure." Sacramento, CA: Concrete Masonry Association of California and Nevada, 1980.

3.17 F. Leonhardt and R. Walther, "Wandartiger Träger." Deutscher Ausschuss für Stahlbeton, Bulletin No. 178. Berlin: Wilhelm Ernst & Sohn, 1966.

3.18 F. Dischinger, "Contribution to the Theory of Wall-Like Girders." *Proceedings*, Volume I, International Association for Bridge and Structural Engineering, 1932.

4.1   "Reinforced Concrete Masonry Columns and Pilasters." Herndon, VA: National Concrete Masonry Association.

4.2   *Uniform Building Code*, 1982 Edition. Whittier, CA: International Conference of Building Officials, 1982.

4.3   "Specification for the Design, Fabrication, and Erection of Structural Steel for Buildings," *Manual of Steel Construction*, Eighth Edition. Chicago: American Institute of Steel Construction, 1980.

4.4   W. M. Simpson, "The Slender Wall Test Program: Part 2." *Masonry Industry Magazine*, March–April 1982, 10.

4.5   R. G. Drysdale, A. A. Hamid, and A. C. Heidebrecht, "Tensile Strength of Concrete Masonry." *Journal of the Structural Division*, ASCE, 105: ST7, 1261–1276, July 1979.

4.6   "Report of Variation of "*b*" or Effective Width in Flexure of a Concrete Block Panel." Los Angeles: Masonry Institute of America, 1966.

4.7   C. S. Whitney, "Plastic Theory of Reinforced Concrete Design." *Transactions*, ASCE, 1942, 107:251–326.

4.8 R. R. Schneider and W. L. Dickey, *Reinforced Masonry Design*. Englewood Cliffs, NJ: Prentice-Hall, 1980.

5.1 ACI Committee 318, *Building Code Requirements for Reinforced Concrete*, ACI 318-83. Detroit, MI: American Concrete Institute, 1983.

5.2 B. Ellingwood, T. V. Galambos, J. G. MacGregor, and C. A. Cornell, "Development of a Probability Based Load Criterion for American National Standard A58," NBS Special Publication 577. Washington, D.C.: National Bureau of Standards, 1980.

5.3 B. Ellingwood, T. V. Galambos, J. G. MacGregor, and C. A. Cornell, "A Probability Based Criterion for Structural Design." *Civil Engineering*, ASCE, July 1981, 74–76.

5.4 F. Y. Yokel, R. G. Mathey, and R. D. Dikkers, "Compressive Strength of Slender Concrete Masonry Walls," Building Science Series 33, Washington, D.C.: National Bureau of Standards, 1970.

5.5 *Notes on ACI 318-77*. Skokie, IL: Portland Cement Association, 1978.

5.6 W. M. Simpson, "The Slender Wall Test Program: Part 2." *Masonry Industry Magazine*, March–April 1982, 10.

5.7 SEAOSC, "Recommended Tilt-Up Wall Design." Los Angeles: Structural Engineers Association of Southern California, June 1979.

5.8 *Uniform Building Code*, 1982 Edition. Whittier, CA: International Conference of Building Officials, 1982.

5.9 R. E. Englekirk and G. C. Hart, "Proposed UBC Code Change." Sacramento, CA: Concrete Masonry Association of California and Nevada, 1980.

5.10 R. Park and T. Paulay, *Reinforced Concrete Structures*. New York: Wiley, 1975.

5.11 S. Chen, P. A. Hidalgo, R. L. Mayes, R. W. Clough, and H. D. McNiven, "Cyclic Loading Tests of Masonry Shear Piers," Volume II, UCB/EERC–78/28. Berkeley: University of California, 1978.

6.1 B. Ellingwood, T. V. Galambos, J. G. MacGregor and C. A. Cornell, "A Probability Based Load Criterion for Structural Design." *Civil Engineering*, ASCE, July 1981.

6.2 G. C. Hart, *Uncertainty Analysis, Loads, and Safety in Structural Engineering*. Englewood Cliffs, N.J.: Prentice-Hall, 1982.

6.3 B. Ellingwood, T. V. Galambos, J. G. MacGregor, and C. A. Cornell, "Development of a Probability Based Load Criterion for American National Standard A58," NBS Special Publication 577. Washington, D.C.: National Bureau of Standards, 1980.

6.4 ACI Committee 318, *Building Code Requirements for Reinforced Concrete*, ACI 318-83. Detroit, MI: American Concrete Institute, 1983.

6.5 "Specification for the Design, Fabrication, and Erection of Structural Steel for Buildings, Part II," *Manual of Steel Construction*, Eighth Edition. New York: American Institute of Steel Construction, 1980.

6.6 "Building Code Requirements for Minimum Design Loads in Buildings and Other Structures," ANSI A58.1-1982. New York: American National Standards Institute, 1982.

6.7 "National Building Code of Canada." Ottawa: National Research Council of Canada, 1977.

7.1 "Ultimate Seismic Design of Reinforced Hollow-Unit Concrete Masonry Shear Walls," Research Report 4039. Whittier, CA: International Conference of Building Officials, April 1983.

7.2 "Strength Design of Shear Walls," Concrete Masonry Association of California and Nevada, 1983.

A.1 B. Ellingwood, T. V. Galambos, J. G. MacGregor, and C. A. Cornell, "Development of a Probability Based Load Criterion for American National Standard A58," NBS Special Publication 577. Washington, D.C.: National Bureau of Standards 1980.

A.2 "Building Code Requirements for Minimum Design Loads in Buildings and Other Structures," ANSI 58.1-1982. New York: American National Standards Institute, 1982.

A.3 B. Ellingwood, J. G. MacGregor, T. V. Galambos, and C. A. Cornell, "Probability Based Load Criteria: Load Factors and Load Combinations." *Journal of the Structural Division*, ASCE, 108:ST5, 978, May 1982.

A.4 ACI Committee 318, *Building Code Requirements for Reinforced Concrete*, ACI 318-77. Detroit, MI: American Concrete Institute, 1977.

A.5 J. G. MacGregor, B. Ellingwood, T. V. Galambos, C. A. Cornell, and S. A. Mirza, "Load and Strength Reduction Factors for the Design of Concrete Structures," under preparation for the American Concrete Institute.

A.6 G. C. Hart, S. C. Huang, and T. A. Sabol, "Calibration of Strength Reduction Factors for Concrete Masonry, *Civil Engineering Systems*, 1:1, September 1983.

A.7 T. V. Galambos and M. K. Ravindra, "Load and Resistance Factor Design." *Journal of the Structural Division*, ASCE ST9, 104:1335, September 1978.

A.8 G. C. Hart, *Uncertainty Analysis, Loads, and Safety in Structural Design.* Englewood Cliffs, N.J.: Prentice Hall 1982.

INDEX